Medical Accidents

Medical Accidents

Edited by

Charles Vincent
Maeve Ennis
Robert J. Audley

University College
London

Oxford New York Tokyo
OXFORD UNIVERSITY PRESS
1993

Oxford University Press, Walton Street, Oxford OX2 6DP

Oxford New York Toronto
Delhi Bombay Calcutta Madras Karachi
Kuala Lumpur Singapore Hong Kong Tokyo
Nairobi Dar es Salaam Cape Town
Melbourne Auckland Madrid
and associated companies in
Berlin Ibadan

Oxford is a trade mark of Oxford University Press

Published in the United States
by Oxford University Press Inc., New York

A catalogue record for this book is available from the British Library

Library of Congress Cataloging in Publication Data
Medical accidents/edited by Charles Vincent, Maeve Ennis, Robert J. Audley. — 1st ed.
Includes bibliographical references and index.
1. Medical errors.   I. Vincent, Charles, psychologist.
II. Ennis, Maeve,   III. Audley, Robert J.
[DNLM: 1. Malpractice.   2. Diagnostic Errors.   3. Iatrogenic
Disease.   4. Insurance, Liability. W 44 M4845 1993]
R729. 8. M43   1993   610—dc 20   93-17624
ISBN 0 19 262289 7

Typeset by Expo Holdings, Malaysia
Printed in Great Britain on acid-free paper by
Bookcraft (Bath) Ltd, Midsomer Norton

Preface

When we began research into medical accidents in 1985 we were astonished to discover that there was virtually no research on the subject. In other fields (e.g. aviation, road and rail travel, nuclear power) accidents were intensively investigated but, in medicine, accidents and errors were seldom discussed publicly. Whilst any doctor we spoke to could provide examples of occasions on which patients had been injured during treatment, or had narrowly avoided serious injury, very few systematic studies had been published. In the last few years, however, there has been a greater willingness to acknowledge and investigate medical accidents and a marked increase in research on their causes and prevention. This growing body of work is, however, reported in the journals of many different disciplines and there is no easy way of obtaining an overall view of this important area. This book attempts to provide this wider perspective.

Several factors have contributed to the growing interest in medical accidents. The rapidly rising rate of litigation in the 1980s, and increasing interest from the media, brought medical accidents to the attention of both doctors and the general public. Systems of complaint and compensation had also been widely criticized and this led to calls for reform from lawyers, doctors, and organizations representing patients. There is now considerable debate about both the kind of system of compensation that should be adopted, and about the extent to which doctors should be held accountable for their actions. Of equal importance has been the growth of medical audit and the wide recognition within the medical profession of the need to constantly monitor quality of care. Recent studies have made it clear that many more patients are injured during their treatment than had been previously acknowledged.

In parallel with these changes in medicine, research workers from several disciplines have been developing methods for the analysis of accidents of all kinds. Theories of error and accident causation have evolved which are applicable across many human activities although they have not as yet been widely used within medicine. These developments have led to a much broader understanding of accident causation, with less focus on the individual who makes an error and more on pre-existing organizational factors that provide the context in which errors occur. An important consequence of this has been the realization that the analysis of accidents has the potential to reveal deep-rooted unsafe features of organizations that are both inefficient and potentially dangerous.

A complete understanding of medical accidents requires an interdisciplinary approach. The views of doctors, psychologists, lawyers, and economists are all represented in this book, and we readily acknowledge that other disciplines are making important contributions. In Section I, the essential features of accident causation and investigation are described and an overview of research on medical accidents is presented. These two chapters describe the methods available to clinicians or researchers who wish to investigate accidents and they also provide an introduction to the three clinical chapters that follow. In each chapter of Section II, a senior doctor discusses the causes and prevention of errors and accidents in his particular specialty. Each clinician combines a review of the research literature in their area with observations from their own clinical and medico-legal experience. In Section III, we consider some important questions that arise in any discussion of the causes of medical accidents. Are we selecting the people most suited to become doctors? Does their education, and their training after qualifying, make their practice sufficiently safe? Do the long hours that doctors work and the stressful nature of their work affect the quality of care they provide? Can an analysis of the processes of medical thinking and decision-making reveal the reasons for faulty decisions, and can it suggest how safer, optimal decisions might be made? These topics are certainly not a complete list of the potential causes of accidents but they are all undoubtedly significant.

In Section IV the emphasis of the book shifts to the consequences for those involved in medical accidents. Litigation has been said to lead to defensive medicine, i.e., medicine practised for legal rather than clinical reasons. This may be one, indirect, effect of medical accidents. Equally important are the immediate effects on the staff involved in an accident. There is almost no research on these but some junior doctors have reported intense guilt and distress after making what were possibly unavoidable mistakes. Even less is known about the effect on patients, although patient organizations have suggested that involvement in a medical accident can be deeply disturbing emotionally as well as physically. An equally distressing feature of some accidents for both patients and doctors is the litigation that may follow.

Procedures for compensation and, to a lesser extent, for dealing with complaints are complex, long-winded, expensive, and exhausting for all concerned. In Section V, existing procedures are reviewed and criticized and options for reform in the areas of both compensation and accountability are reviewed. Although no-fault compensation has many attractions, it is now clear, firstly, that there are several possible varieties of no-fault compensation and, secondly, that no-fault schemes on their own do not address all the concerns of patients: particularly their wish to see changes in medical practice follow from the results of their experiences. In Section VI, we draw on all the previous chapters to present an overview of themes that recur throughout the book. For example, the theme of the organization as a source of accidents surfaces in a number of chapters and from several different perspectives.

The editor of this book have had very definite aims. We hoped it would stimulate interest and further research in this important and complex area: there are certainly theoretical questions in psychology and law; and questions relating to the conduct and methodology of medical audit that can be examined in the context of medical accidents. Above all, we hope that the book will be of use to both organizations and individuals who are concerned with understanding and preventing accidents and with helping those involved. We believe the book shows how errors and accidents in medicine can be studied in a way that will enable clinicians to develop their own research into accidents, and to minimize the effects of the errors and accidents that will, inevitably, still occur. We also wish to draw attention to the problems associated with procedures for compensation and accountability and promote the introduction of much needed reforms.

Finally, we should like to say something about what this book is not. We hope it is already clear that it is not in any sense an attack on the integrity of the medical profession nor on the practice of medicine, any more than a study of road accidents is an attack on driving. Our own research programme has always depended critically on collaboration with doctors. We hope this book will be seen by them as a contribution towards their own goal of improving the care of patients.

London   C. V.
June 1993   M. E.
                                                                        R. J. A.

Acknowledgements

Our principal thanks must go to those who have contributed chapters to this book, many of whom have also been collaborators and advisors on our research programme. We wish to express our gratitude to Laura Olivieri for her support, assistance, and constant good humour during the writing of the book and the last year of the project. We also gratefully acknowledge the assistance of Margaret Beringer in the preparation of the manuscript.

Our own research and this book owe their origins to a suggestion to Professor Audley by Desmond Laurence, then Professor of Clinical Pharmacology at UCL, that the medical profession should turn to psychologists to help them in the study of medical mishaps. An early result of this was the funding of a research project by the Medical Protection Society. In the course of this and later projects we received particular help from the following members of the Council and the staff of the Society: Sir John Ellis, Peter Ford, Alan Brown, Roy Palmer, John Hickey, Professor Jack Stevens, Paul Miller, and Christopher Orr (who is also the author of a chapter).

Much of what is reported in this book has depended on the generous collaboration of not only several of our authors but also many clinicians (both senior and junior), the patients in their care, and lawyers. It is not possible to mention them all by name but we especially wish to acknowledge the major contribution of Peter Driscoll.

Finally, we wish to acknowledge the intellectual support given by many of our colleagues in the Department of Psychology at University College London.

Contents

Contributors

Robert J. Audley Head of Psychology Department and Vice-Provost, University College London, Gower Street, London WC1E 6BT

Jack Dowie Senior Lecturer, Department of Social Science, Open University, Walton Hall, Milton Keynes MK7 6AA

James O. Drife Professor of Obstetrics and Gynaecology, Academic Unit of Obstetrics and Gynaecology, University of Leeds, D Floor, Clarendon Wing, Belmont Grove, Leeds LS2 9NS

Maeve Ennis Lecturer in Health Psychology, formally Medical Protection Society Research Fellow, Department of Psychology, University College London, Gower Street, London WC1E 6BT

Paul Fenn Research Fellow, Centre for Socio-Legal Studies, Wolfson College, Oxford OX2 6UD

Jenny Firth-Cozens Lecturer in Psychology, Department of Psychology, and Associate of Nuffield Institute for Health, University of Leeds, Leeds LS2 9JT

J. Gedis Grudzinskas Professor of Obstetrics and Gynaecology, Academic Department of Obstetrics and Gynaecology, The Royal London Hospital Medical College, Whitechapel Road, London E1 1BB

Chris McManus Senior Lecturer in Psychology, Department of Psychology, University College London, Gower Street, London WC1E 6BT and Academic Department of Psychiatry, St Mary's Hospital Medical School, Praed Street, London W2 1NY

Christopher Orr Regional Medico Legal Adviser, East Anglian Regional Health Authority, Union Lane, Chesterton, Cambridge CB4 1RF

James T. Reason Professor of Psychology, Department of Psychology, University of Manchester, Oxford Road, Manchester M13 9PL

Ian H. Robertson Clinical Neuropsychologist, MRC Applied Psychology Unit, 15 Chaucer Road, Cambridge CB2 2EF

John H. Scurr Senior Lecturer and Consultant Surgeon, Department of Surgical Studies, Middlesex Hospital, Mortimer Street, London W1 8AA

Arnold Simanowitz Executive Director, Action for Victims of Medical Accidents, Bank Chambers, 1 London Road, Forest Hill, London SE23 3TP

Charles Vincent Lecturer in Psychology, Department of Psychology, University College London, Gower Street, London WC1E 6BT and Academic Department of Psychiatry, St Mary's Hospital Medical School, Praed Street, London W2 1NY

Michael Wilson Consultant Anaesthetist, Department of Anaesthetics, Royal United Hospital, Combe Park, Bath BA1 3NG

1

The human factor in medical accidents

JAMES T. REASON

Introduction

Over the past few years, there has been a noticeable spirit of *glasnost* within the medical profession concerning the role played by human error in the causation of medical accidents. This has been particularly apparent in the field of anaesthetics (Cooper *et al.*, 1978, 1984; Cook and McDonald, 1988; Gaba, 1989) and in 'high-tech' facilities such as intensive care units (Galer and Yap, 1980; Gopher *et al.*, 1989).[1]

One of the most obvious signs of this new openness has been the recent involvement of human factors specialists in studies of patient safety. This has brought at least two obvious benefits. The first is mainly methodological, permitting techniques like critical incident analysis and event reporting programmes, initially developed in the field of aviation, to be applied to studying the medical accident process. Second, the results of these and other investigations have clearly shown that medical mishaps share many important causal similarities with the breakdown of other complex socio-technical systems.

In their important discussion of these common features, Woods and Cook (1991) argue that there is a 'deep structure' to human–system interaction that extends beyond domain boundaries. They conclude with the following statement:

Applying the safety lessons from aircraft to nuclear power and from nuclear power to anaesthesiology and from anaesthesiology to railroad operation cannot be based on [the] surface details of these domains. Because disasters and near misses are rare in each domain it seems reasonable to try to elucidate the deep structure of the domains in [a] common language so as to fill in our model of human performance. Applying safety lessons from one domain to another is contingent on developing empirically grounded and testable models of deep structure which demonstrate that what appears similar in a vague way suggested by analogy is in fact similar at a more concrete, objective and deeper level [p. 16].

---

[1] Other contributors to this book will consider the extent to which human failures are implicated in medical accidents. For our purposes, it is sufficient to assert that their significance is not in dispute.

The present chapter pursues this challenge by exploring whether a model of accident causation, developed to accommodate the human contribution to the breakdown of a wide range of complex technological systems (Reason, 1989, 1990, 1991), has some utility in the context of medical accidents. As with any view of accident causation, utility must be judged by practical rather than purely scientific criteria. Does the model help us to identify the root causes of past accidents? Does it provide a principled basis for developing more effective methods of accident prevention?

Active and latent human failures

Human decisions and actions play a major part in nearly all accidents. This is not so much a question of incompetence or irresponsibility as of opportunity. All potentially hazardous technologies are designed, built, operated, and maintained by human beings. Catastrophic breakdowns of complex human–machine systems arise from the combined effects of human failures in all of these activities.

Humans contribute to accidents in two ways: through active failures and latent failures. These two categories are distinguished both by the time taken for the failure to have a negative impact upon the safety of the system, and by the kind of person responsible.

Active failures are unsafe acts committed by those at the 'sharp end' of the system (pilots, air traffic controllers, ships' crews, train drivers, control room operators, maintenance crews, anaesthetists, surgeons, nurses, and the like). They are the people at the human–system interface whose actions can, and sometimes do, have immediate adverse consequences. Quite often, these unsafe acts involve the circumvention or disabling of safety devices designed to protect the system against serious breakdown.

Latent failures arise from fallible decisions, usually taken within the higher echelons of the organization or within society at large. Their damaging consequences may lie dormant for a long time, becoming evident only when they combine with local triggering factors (i.e., active failures, technical faults, atypical system states, and so on) to breach the system's defences. Most often, the people primarily responsible for the commission of these latent failures are separated in both time and distance from the hazardous workplace.

Until quite recently, it was usual for most accident investigations to focus upon the active rather than the latent human failures. Most investigators saw their task as identifying those people or equipment items that were immediately responsible for the system breakdown, and then specifying a list of local repairs that would prevent a recurrence of that particular accident sequence. In general, they had neither the resources nor the training to track down the long-standing organizational causes.

It is difficult to identify the precise moment of change, but in Britain a significant landmark was Mr Justice Sheen's judgement regarding the capsize of the car ferry, *Herald of Free Enterprise*. In addition to identifying active failures on the part of the Master and members of the crew, he went on to state that: '... a full investigation of the circumstances of the disaster leads inexorably to the conclusion that the underlying cardinal faults lay higher up the Company.' (Sheen, 1987). Similar judgements were subsequently made in regard to the causes of the King's Cross Underground fire (Fennell, 1988), the Hillsborough Stadium disaster, the Clapham Junction rail collision (Hidden, 1989), and the Piper Alpha oil platform explosion (Cullen, 1990).

In these cases, as in many others, the front-line personnel were the inheritors rather than the instigators of disaster. Their part was to create the conditions under which the latent system failures could manifest themselves.

Three universal accident ingredients

There are three reasons why the possibility of an accident occurring to some hazardous human activity can never be wholly discounted.

1. All human beings, regardless of their skills, abilities, and specialist knowledge, make fallible decisions and commit unsafe acts. This human propensity for committing errors and violating safety procedures can be moderated by selection, training, well-designed equipment, and good management, but it can never be entirely eliminated.
2. No matter how well designed, constructed, operated, and maintained they may be, all man-made systems possess latent failures in some degree. These failures are analogous to resident pathogens in the human body that combine with local triggering factors (i.e., life stresses, toxic chemicals, and the like) to overcome the immune system and produce disease. Like cancers and cardiovascular disorders, disasters in well-defended systems do not arise from single causes. They occur because of the adverse conjunction of several factors, each necessary but none sufficient to breach the defences. As in the case of the human body, no technological system can ever be entirely free of pathogens.
3. All human endeavours involve some measure of risk. In many cases, the local hazards are well understood and can be guarded against by a variety of technical or procedural counter-measures. But no one can foresee all the possible accident scenarios, so there will always be chinks in this protective armour.

These three ubiquitous accident ingredients reveal something important about the nature of the 'safety war'. They tell us that the fight against accidents is not like a conventional war in which a decisive victory can be followed by a long period of relative peace. There will be no Waterloos. Rather, the pursuit of safety has the character of a guerilla conflict in which rigidity, complacency, and an over-reliance on technical 'fixes' are certain routes to defeat. Maintaining safety within acceptable limits requires constant vigilance and chronic unease, both difficult to sustain when there are always many other demands upon limited human and financial resources.

Above all, however, we must struggle to understand the true nature of the enemy — in this case, the insidious accumulation of latent organizational failures in otherwise well-defended systems. It will be argued here that the best chance of minimizing accidents is by identifying and correcting these delayed-action failures *before* they combine with local triggers to breach or circumvent the system's defences.

Measuring safety: the accident paradox

Many organizations use accident data as an index of the relative safety of their constituent parts or subsystems. But accidents, by themselves, are a poor indication of general 'safety health'. Only if a system had complete control over all accident-causing factors could its accident history provide a valid measure of its safety. Only then would its negative outcome data be a direct reflection of shortcomings in its safety management. But, as argued above, no organization can ever achieve this complete control. The local hazards can be moderated, but never eliminated. Latent system failures will always be present in some degree and the probability of their chance conjunction with local triggers and unsafe acts is always greater than zero. The large stochastic element in accident causation is such that 'safe' organizations can still have bad accidents, while relatively 'unsafe' ones can escape them for long periods.

How can we resolve this accident paradox? One way is to recognize that safety has two faces. The positive face of safety (i.e., intrinsic resistance to chance combinations of hazards, unsafe acts, and technical failures), like good health, is difficult to pin down and even harder to measure. By comparison, the negative face of safety is all too clearly signalled by near misses, injuries, and fatalities that lend themselves to close analysis and quantification.

However, the data provided by accident and incident reporting systems, while essential for understanding the causes of past mishaps, are both too little and too late to support measures directed at enhancing a system's intrinsic safety health. For this purpose, we need to monitor the system's 'vital signs' on a regular basis (i.e., indices relating to the quality of management, equipment design and construction, conditions of work, safety procedures, communications, maintenance, and so on). Only these factors lie within the organization's direct sphere of influence, and only their improvement constitute achievable safety goals. Accidents, by their nature, are not directly controllable, and this is particularly so when they have been reduced by good practice to extremely rare events (as in commercial aviation or medicine).

The anatomy of an organizational accident

The rare accidents that still occur within systems possessing a wide variety of technical and procedural safeguards (such as operating theatres and intensive care units) have been termed *organizational accidents* (Reason, 1990). These are mishaps that arise not from single errors or isolated component breakdowns, but from the insidious accumulation of delayed action failures lying mainly within the managerial and organizational spheres. Such latent failures may subsequently combine with active failures and local triggering factors to penetrate or bypass the system defences. These residual problems do not belong exclusively to either the machine or the human domains. They emerge from complex and still little-understood interactions between the social and technical aspects of the total system.

What sorts of ideas are necessary for understanding the causes of organizational accidents? It is argued here that a theory describing such accident sequences requires four basic concepts: organizational processes, task and environmental conditions, individual unsafe acts, and failed defences. The relationships between these theoretical elements are summarized in Fig. 1.1.

The direction of causality in Fig. 1.1 is from left to right. Fallible decisions associated with the generic processes of all technical organizations (i.e., setting goals, organizing, managing, communicating, designing, building, operating, and maintaining) seed 'resident pathogens' (or latent failures)

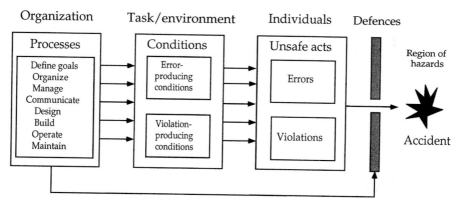

Figure 1.1   Relationships between the theoretical elements of organizational accidents.

within the system. These are then transmitted along departmental pathways to the workplace, where they give rise to the task and environmental conditions likely to promote unsafe acts. Many unsafe acts may be committed, but very few of them are likely to occur in the presence of absent or failed defences, though such actions can often involve the bypassing or disabling of established safeguards. But the system's defences may already be deficient due to latent failures; such a possibility is shown in Fig. 1.1 by the arrow connecting organizational processes directly to defences.

Each of these elements is considered in further detail below, starting with the failed defences and working backwards to the root organizational processes.

DEFENCES

Defences are measures designed to protect against hazards and to mitigate the consequences of equipment or human failures. Defences comprise both technical and human elements. As mentioned earlier, many technologies are now equipped with a wide range of computerized safety systems that can act automatically to restore the system to a safe state. In some of these 'high-tech' installations (i.e., nuclear power and certain chemical process plants, intensive care units, and so on), the primary function of the human operator is to diagnose and correct faults that lie outside the scope of these engineered defences. Unfortunately, this is not something that the operator's abilities, experience, or training can guarantee (Bainbridge, 1987), despite the fact that human beings still remain the only information-processing devices capable of coping with entirely novel situations.

Defences serve several functions, each of them dependent upon an accurate awareness of the nature of the hazards. Any one of these functions may fail singly or in combination.

Protection: To provide a barrier between the hazards and the potential victims under normal operating conditions.

Detection: To detect and identify the presence of an off-normal condition or hazardous substance.

Warning: To signal the presence and the nature of the hazard to all those likely to be exposed to its dangers.

Recovery: To restore the system to a safe state as quickly as possible.

Containment: To restrict the spread of the hazard in the event of a failure in any or all of the prior defensive functions.

Escape: To ensure the safe evacuation of all potential victims should the spread of the hazard become uncontrolled.

UNSAFE ACTS

Unsafe acts divide into two distinct groups: *errors* and *violations*. All involve deviations but they differ with regard to the nature of these departures.

Errors may be of two basic kinds (Reason, 1990): (a) *attentional slips* and *memory lapses,* involving the unintended deviation of actions from what may be a perfectly good plan; and (b) *mistakes,* where the actions follow the plan but the plan deviates from some adequate path to the desired goal.

Mistakes fall into two groups: (i) *rule-based mistakes,* in which the individual encounters some relatively familiar problem, but applies the wrong prepackaged solution (either the misapplication of a good rule, or the application of a bad rule); and (ii) *knowledge-based mistakes,* in which the individual encounters a novel situation for which his/her training has not provided some rule-based solution. The consequence is that he/she has to use on-line reasoning based upon an incomplete or incorrect mental model of the problem.

Violations involve deliberate deviations from some regulated code of practice or procedure. They divide into four types: (a) routine violations, involving short cuts between task-related points; (b) optimizing violations, in which the individual seeks to optimize some goal other than safety (alleviate boredom, for instance); (c) exceptional violations, one-off breaches of regulations seemingly dictated by unusual circumstances; and (d) deliberate sabotage. It must be emphasized that not all violations cause accidents; sometimes they save lives and earn commendations. The varieties of unsafe acts are summarized in Fig 1.2.

CONDITIONS PROMOTING UNSAFE ACTS

Each type of activity has its own nominal error probability. For example, carrying out a totally novel task with no clear idea of the likely consequences (i.e., knowledge-based processing) has a basic error probability of 0.75. At the other extreme, a highly familiar, routine task performed by a well-motivated and competent workforce has an error probability of 0.0005. But there are certain conditions both of the individual and of his/her immediate environment that will increase these nominal error probabilities. These are summarized in Table 1.1 (Data from Williams, 1988). Here, the error-producing conditions are ranked in the order of their known effects and the numbers in parentheses indicate the risk factor (i.e., the amount by which the nominal error rates should be multiplied under the worst conditions).

For convenience, we can reduce these many error-producing conditions to seven broad categories: high workload; inadequate knowledge, ability, or

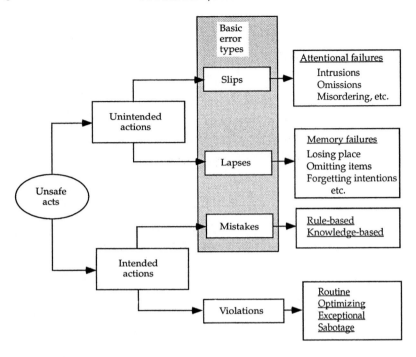

Figure 1.2 Varieties of unsafe acts.

Table 1.1 Error-producing conditions

Unfamiliarity with the task (×17)
Time shortage (×11)
Poor signal:noise ratio (×10)
Poor human–system interface (×8)
Designer-user mismatch (×8)
Information overload (×6)
Negative transfer between tasks (×5)
Misperception of risk (×4)
Poor feedback from system (×4)
Inexperience (not lack of training) (×3)
Poor instructions or procedures (×3)
Inadequate checking (×3)
Educational mismatch of person with task (×2)
Disturbed sleep patterns (×1.6)
Hostile environment (×1.2)
Monotony and boredom (×1.1)

Table 1.2   Violation-producing conditions

Manifest lack of organizational safety culture
Conflict between management and staff
Poor morale
Poor supervision and checking
Group norms condoning violations
Misperception of hazards
Perceived lack of management care and concern
Little élan or pride in work
A macho culture that encourages risk-taking
Beliefs that bad outcomes won't happen
Low self-esteem
Learned helplessness
Perceived license to bend rules
Ambiguous or apparently meaningless rules

---

experience; poor interface design; inadequate supervision or instruction; a stressful environment; mental state (fatigue, boredom, and so on); and change.

Departures from routine and changes in the circumstances in which actions are normally performed constitute a major factor in the making of absent-minded slips of action (Reason and Mycielska, 1982).

Compared to error-producing conditions, the factors that promote violations are less well understood. It is not possible to rank their relative effects as we have for errors. However, we can make an informed guess at the nature of these violation-producing conditions. They are shown in Table 1.2, although in no particular order of effect.

Again, we can reduce this potentially very long list to a limited number of general categories: lack of safety culture, lack of concern, poor morale, violation-condoning norms, macho attitudes and poorly expressed rules.

ORGANIZATIONAL PROCESSES

All technological organizations, whether they be hospitals, transport systems, or process plants, share a number of common processes. These have been discussed at length elsewhere (Reason, 1989, 1990, 1991), but can be summarized as follows: goal setting, organizing, communicating, managing, designing, building, operating, and maintaining. Fallible decisions made in regard to these processes give rise to the following *organizational failure types:* incompatible goals, structural (organizational) deficiencies, inadequate communications, poor planning and scheduling, inadequate control and monitoring, design failures, deficient training, and inadequate maintenance management.

It must be emphasized that these are not the only possible failure types, but they are the ones that feature most commonly among the root causes of organizational accidents (see Reason, 1990). In addition, regular assessments of the state of these failure types (or 'vital signs') would provide a reasonably comprehensive indication of an organization's current 'safety health'.

A case study illustration

The best way of putting flesh onto these bare theoretical bones is to consider a case study example. The one considered below is drawn from the field of anaesthetics and has been presented by Eagle *et al.* (1992) specifically to illustrate the arguments set out above.

A 72-year-old man was listed to have a cystoscopy performed under local anaesthetic. Because the patient was booked for a local anaesthetic, he was not assessed by an anaesthetist beforehand.

On the morning in question, the attending urologist discovered that he was double-booked in the operating theatre, so he asked a urologist colleague to carry out the procedure in his stead. This colleague, who had not seen the patient, agreed but insisted that the procedure should be carried out under a general anaesthetic.

The second urologist contacted the anaesthetic co-ordinator who found an available anaesthetist, but this anaesthetist was not told of the original booking under local anaesthesia, nor that the urologist performing the procedure was unfamiliar with the patient. The anaesthetist merely understood that the patient was an urgent addition to the operating theatre list.

The anaesthetist first saw the patient in the operating theatre where he was found to be belligerent, confused, and unable to give a coherent history. The nursing notes showed that the patient had fasted for 24 hours, but his chart revealed several complications. These included progressive mental deterioration, metastatic cancer in the lungs and liver, renal insufficiency, and anaemia.

Since the patient refused to be moved to the cystoscopy table, the anaesthetist decided to induce anaesthesia with the patient in his bed. Routine monitors were attached and Thiopentone (150 mg) and alfentanil (500 μg) administered intravenously. The patient lost consciousness, but then regurgitated more than two litres of fluid and undigested food. He was immediately turned on his side and the vomitus sucked from his mouth. The patient's lungs were ventilated with 100 per cent oxygen and intravenous fluids were given. Investigation showed large quantities of fluid in the patient's bronchi. He was transferred to the Intensive Care Unit and died six days later.

FAILED DEFENCES

The possibility of vomiting under anaesthesia (carrying the risk of death through aspiration) is a well-understood surgical hazard. It is normally defended against by preoperative fasting, by vigilant nursing, and by a detailed assessment of the patient by the anaesthetist both on the ward and during the preoperative preparation. The concatenation of a number of both active and latent failures contrived to breach these 'defences-in-depth' in this instance. As a result, the protection, detection, and warning functions of these procedural defences were invalidated and the recovery and containment functions, though promptly applied, proved ineffective.

UNSAFE ACTS

Active failures were committed by two people. They are listed below in temporal order.

1. The first urologist committed a rule-based mistake when booking the patient for cystoscopy under a local anaesthetic. By applying the rule of thumb that 'elderly patients are given a local anaesthetic for cystoscopy', the attending urologist overlooked the counter-indications, namely that this was a confused and combative patient, and selected the wrong form of anaesthetic.

2. The anaesthetist similarly made a series of rule-based mistakes in assuming the fasting status indicated on the nursing notes was correct, in assuming that all the relevant data were available on the chart, and in deciding to proceed with the anaesthetic despite having performed a fairly cursory preoperative assessment on a patient with several medical problems.

ERROR-ENFORCING CONDITIONS

The anaesthetist, in particular, was the unknowing focus of a number of resident pathogens. Failures in the operating theatre scheduling system caused the first urologist to be double-booked. Breakdowns in communication caused the anaesthetist to be unaware of the substitution of a second urologist and of the original booking for the procedure to be performed under local anaesthetic. This opened a serious breach in the preoperative defences, and resulted in the patient not being assessed by an anaesthetist on the evening before surgery.

In addition, a post-event review of the patient's chart revealed that an episode of projectile vomiting had occurred at 4 a.m. on the morning of the procedure. This information was recorded by a nurse in the computerized record-keeping system, but had not been printed out and appended to the chart due to lags in the system. Furthermore, no terminal was available in the operating theatre, so the anaesthetist could not, in any case, have checked the computerized record for recent updates. The anaesthetist was unaware of the delay in producing hard copies of the patient's record and, most importantly, was thus ignorant of the vomiting incident which would have created strong suspicion as to the 24-hour fasting claim.

ORGANIZATIONAL FAILURES

Several of the previously listed organizational failure types featured prominently among the root causes of this accident. First and most obviously, there were significant *communication failures* at a number of points in the system. Second, there were deficiencies in the operating theatre *management system* that allowed the first urologist to be double-booked. Third, the *design and operation* of the computerized record-keeping system allowed crucial information to be denied to those at the 'sharp end'. It is also likely that the absence of a terminal in the operating theatre was the result of an *administrative conflict* between the pursuit of cost-saving and safety goals.

This accident illustrates very clearly how a number of active and latent failures can combine to create a major gap in the system's defences. In such a multiply-defended system, no one of these failures would have been sufficient to cause the patient's death; each provided a necessary link in the causal chain.

Consequences of an 'organizational' analysis

What is to be gained by pursuing accident causes beyond the active failures that are the immediate triggers for a mishap? Some might argue that this merely blurs the issue of professional liability by shifting the 'blame' further 'upstream' within the system.

The first point is that although litigious western societies are apparently obsessed with identifying whom to sue, such a 'blame culture' is of little or no use when it comes to understanding the complex interacting causes of medical mishaps, or in identifying the appropriate remedial measures. The present organizational perspective portrays the human agents of system breakdown as victims rather than villains, as the inheritors rather than the instigators of accident sequences that have their beginnings much earlier than the immediately precipitating events.

The second benefit is one of emphasis. While much progress has been made within the medical profession in their study of mishaps, the investigative spotlight still shines most brightly on those at the 'sharp end': the surgeons, physicians, anaesthetists, radiologists, and nurses who are in direct contact with patients. Such a focus is natural in a field peopled by highly trained professionals whose ethos requires the acceptance of personal responsibility. But there is a danger of becoming stuck in the 'human error' groove, something that has happened in other fields, especially in aviation, where the term 'pilot error' has become a catchall category for most non-technical accident causes. Neither doctors nor pilots work alone, nor do they commit errors within some isolated personal space. Rather, they work in complex, tightly coupled organizational settings whose processes and their inevitable failings need to be understood if accident rates are to be improved beyond (or even maintained at) their present exceedingly low levels.

The third and most important benefit, however, is that an organizational analysis of accident causation provides the only sound basis for effective safety management. Westrum (1988) has classified organizations into three categories: pathological, calculative, and generative, according to their dominant mode of response to hazards. Pathological organizations react by *denying* the dangers; calculative organizations respond reactively by applying *local repairs* after the event (i.e., disciplining those who commit active failures and supplying engineering 'fixes' to prevent the recurrence of specific technical failures); generative organizations, on the other hand, see each event or mishap as a spur to *reforming* the system as a whole. In part this has to do with achieving a flexible and adaptive organizational structure that can cope with both routine and high-tempo modes of operation (La Porte and Consolini, 1988). But generative organizations also function proactively to identify and remove latent failures *before* they combine with local triggers to breach or bypass the system's defences.

As pointed out earlier, fallibility is an ineradicable part of the human condition. And for good reasons: recurrent error forms have their origins in highly adaptive cognitive processes, and are the debit side of an otherwise formidable credit balance (see Reason, 1990). They can be moderated by good training, appropriate procedures, well-designed human-machine interfaces and the like, but they can never be entirely eliminated altogether. Addressing only the active failures and not their environmental and systemic causes is a tokenistic approach (Hudson, personal communication). Errors are the token 'children' of the 'parent' organizational failure types. They are like the dragon's teeth of the fable: remove one and another springs up in its place. Failing to look beyond the errors and the individuals who commit them to their source types will freeze an organization, at worst, in the pathological mode or, at best, in the 'firefighting' or calculative mode of safety management.

So what are the solutions? They are essentially twofold. First, we must design systems that acknowledge human fallibility, recognize its varieties, and are forgiving of unsafe acts. As Norman (1988) has pointed out, many human–machine systems are unwittingly designed to promote avoidable errors rather than to prevent them. Second, we must make the system as a whole more transparent to the presence of its inevitable 'resident pathogens' or latent failures. Some techniques for doing this are outlined below in the concluding section.

Remedial measures

The remedial implications of the organizational approach to accident causation are summarized in Fig. 1.3. As mentioned earlier, the key to enhanced safety lies in appreciating what is manageable and what is not. Accidents, by their nature, are not directly controllable: too much of their causal variance lies outside the organization's sphere of influence. The same, to a large degree, is true of unsafe acts. They can be moderated, but they can never be eliminated. Similarly, the organization can only improve and extend its defences against hazards: it cannot remove these task-related dangers completely.

The causal factors most amenable to regular monitoring and control are the organizational failure types considered above. They are the manageable factors that determine an organization's overall 'safety health' and, as a consequence, they are the proper focus of effective safety management. A number of techniques now exist for carrying out regular audits of organizational safety health on a proactive basis: the International Safety Rating Scale (International Loss Control Institute, 1988), Shell's TRIPOD (Reason *et al.*1988), and Technica's MANAGER (Pitblado *et al.* 1990). Each one assesses system 'health' along a number of dimensions, corresponding approximately to the organizational failure types.

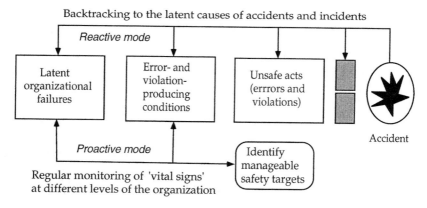

Figure 1.3   Remedial measures for organizational accidents.

Such instruments allow organizations first to identify their principal systemic weaknesses and then to set themselves safety goals that are attainable within the scope of their limited resources. One approach is to target the worst two of these organizational failure types for immediate improvement at each of the regular audits. The same techniques employed to make the initial diagnosis can also be used on subsequent assessments to monitor the progress of remedial efforts and to identify new failure types for correction. Given the unattainability of zero accidents, the best that an organization can achieve is a state of maximum resistance against the disaster-producing factors and then to sustain it for as long as possible. Maintaining an optimal level safety is thus analogous to a long-term fitness programme.

The reactive mode of application is intended to ensure that an organization learns the right lessons from its past misfortunes. To date, it has taken lengthy and expensive public inquiries to reveal the organizational failures lying at the heart of major accidents. However, as shown in our case study example, a structured theoretical framework makes it possible to track back in a principled manner from the accident events to the underlying organizational causes. At each step, causal connections can be made between the factors identified and their origins at the next higher level. In this way, it is possible to carry out far-reaching *organizational reforms*, rather than the *local repairs* that were so often the end result of earlier accident investigations.

References

Bainbridge, L. (1987). The ironies of automation. In *New technology and human error* (ed. J. Rasmussen, K. Duncan, and J. Leplat). Wiley, London.

Cook, R. I. and McDonald, J. S. (1988). Cognitive tunnel vision in the operating room: Analyses of cases using frame model. *Anesthesiology* 69, A497.

Cooper, J. B., Newbower, R. S., and Long, C. D. (1978). Preventable anesthesia mishaps: A study of human factors. *Anesthesiology* 49, 399–406.

Cooper, J. B., Newbower, R. S., and Kitz, R. J. (1984). An analyses of major errors and equipment failures in anesthesia management: Considerations for prevention and detection. *Anesthesiology* 60, 34–42.

Cullen, The Hon Lord. (1990). *The public inquiry into the Piper Alpha disaster.* Department of Energy. HMSO, London.

Eagle, C. J., Davies, J. M., and Reason, J. T. (1992). Accident analysis of large scale technological disasters applied to an anaesthetic complication. *Canadian Journal of Anaesthesia* 39, 118–122.

Fennell, D. (1988). *Investigation into the King's Cross underground fire.* Department of Transport. HMSO, London.

Gaba, D. M. (1989). Human error in anesthetic mishaps. *International Anesthesiology Clinics* 27, 137–47.

Galer, I. A. and Yap, B. L. (1980). Ergonomics in intensive care, applying human factors data to the design and evaluation of patient monitoring systems. *Ergonomics* **23**, 763–79.

Gopher, D., Olin, M., Badih, Y., Cohen, G., and Donchin, Y. (1989). The nature and causes of human errors in a medical intensive care unit. *Proceedings of the Human Factors Society 33rd Annual Meeting*, 956–960.

Hidden, A. (1989). *Investigation into the Clapham Junction railway accident*. Department of Transport. HMSO, London.

International Loss Control Institute (1988). *International safety rating system*. ILCI, Longavill, Georgiao.

La Porte, T. R. and Consolini, P. M. (1988). Working practice but not in theory: Theoretical challenges of high-reliability organizations. *Proceedings of the Annual Meeting of the American Political Science Association*. APSA, Washington, DC.

Norman, D. A. (1988). *The psychology of everyday things*. Basic Books, New York.

Pitblado, R. M., Williams, J. C., and Stater, D. H. (1990). Quantitative assessment of process safety programmes. *Plant/Operations Progress* **9**, 169–75.

Reason, J. (1989). The contribution of latent human failures to the breakdown of complex systems. *Philosophical Transactions of the Royal Society (London)*, *B***327**, 475–84.

Reason, J. (1990). *Human errors*. Cambridge University Press, New York.

Reason, J. (1991). *Identifying the latent causes of aircraft accidents before and after the event*. Paper presented to the International Society of Air Safety Investigators 22nd Annual Seminar, Canberra (unpublished conference proceedings).

Reason, J. and Mycielska, K. (1982). *Absent-minded? The psychology of mental lapses and everyday errors*. Prentice-Hall, Englewood Cliffs, New Jersey.

Reason, J., Shotton, R., Wagenaar, W. A., Hudson, P., and Groeneweg, J. (1988). *TRIPOD: A principled basis for safer operations*. Shell Internationale Petroleum Maatschappij, The Hague.

Sheen, Mr Justice. (1987). *MV Herald of Free Enterprise. Report of Court No.8074. Formal investigation*. Department of Transport. HMSO, London.

Westrum, R. (1988). *Organizational and inter-organizational thought*: World Bank Workshop on Safety Control and Risk Management. Washington, DC,

Williams, J. (1988). A data-based method for assessing and reducing human error to improve operational performance. In *1988 IEEE Fourth Conference on Human Factors and Power Plants* (ed. W. Hagen). Institute of Electrical and Electronic Engineers, New York.

Woods, D. and Cook, R. I. (1991). *Dynamic problem solving in anaesthesiology: Expertise and error*. Position paper for conference on: Human Error in Anesthesia, Monterey, California (unpublished conference proceedings).

2

The study of errors and accidents in medicine

CHARLES VINCENT

This chapter presents a broad overview of research on medical accidents. It begins by summarizing what is known about their nature and incidence and then examines a range of related studies including occurrence screening, closed claims analyses, critical incident studies, confidential enquiries, and medical audit. In each case the focus is on the methods employed rather than the results and possible clinical implications, as these are generally discussed in later chapters. There is also the question of whether studying accidents is a useful approach to improving quality of care. In a final section it is argued that the study of errors and accidents is an essential part of any comprehensive quality initiative and that it shows us that we are often mistaken in our assumptions both about what is happening during routine care and about the reasons things go wrong when they do.

Two factors have particularly contributed to a growing interest in medical accidents and a need to understand their causes. The first is the rapidly rising rate of litigation, payments, and defence society subscriptions during the 1980s (Delamothe, 1989). This certainly led to increased worries over litigation, and some changes in clinical practice (see Chapter 11), but it does not seem to have stimulated much research. Much more important has been the growth of medical audit, jointly fostered by government initiatives and increasing concerns within the health care professions about the quality of care. The rationale of audit is that regular, critical, and systematic analysis of the quality of care will allow clinicians to change their routine practice to improve patient care (Moss, 1992). Progress in these aims has been slow and difficult but audit is certain to continue and become increasingly important. While few audits have examined medical accidents, there is at least now a climate in which to examine errors and accidental injuries to patients is seen as a sensible and acceptable activity.

Some commonly used terms: adverse events, accidents, and negligence

The precise definition of all these terms is difficult, and there is certainly room for disagreement about certain kinds of events. However, the establishment of some working definitions will be attempted. The Harvard Medical Practice Study (1989, described below) defined an adverse event as 'an unintended injury that was caused by medical management and that resulted in measurable disability'. Some of these events might be unforesee-able consequences of certain types of treatment, whose prevention might depend on further advances in medical knowledge. Others are due to human error, and hence potentially preventable. These are considered, in this chapter, to be medical accidents. The varieties of human error that may contribute to these accidents are discussed below and, in more detail, in the previous chapter.

There is a tendency to automatically link medical accidents with malpractice and negligence — occasions where the care given both falls below an acceptable standard and causes an injury to the patient. However, the purpose of studying errors and accidents in medicine is not to blame people and seldom to identify incompetent or negligent individuals. Very few errors, and only a proportion of accidents, involve negligence and the reasons for studying them concern the much broader issue of improving the overall quality of care given to patients.

Nature and frequency of medical accidents

Official statistics are collected on many forms of accidents, but little is known about the overall incidence of medical accidents in the UK. The protection societies have information on the annual rate of claims and complaints, but this is only a fraction of the total number of accidents. Only two studies, both carried out in the USA, have attempted to assess the nature and frequency of medical accidents. The first, conducted in California in 1974, estimated that 4.7 per cent of hospital admissions led to injuries, with 0.8 per cent being due to negligence (Mills, 1978). The most important study in this area, however, has been the extraordinarily thorough and comprehensive Harvard Medical Practice Study (1989). The primary motivation for the Harvard study was the need to examine alternatives to the tort system of damages (see Chapter 12). The system is widely seen as unsatisfactory for doctors, patients, and even for lawyers (Leape *et al.*, 1991). Progress in thinking about possible alternatives (particularly no-fault compensation) was hampered by a lack of accurate information on all the critical issues. The Harvard study asked four main questions: (a) How frequent is medical injury among hospitalized patients, and what fraction of these injuries result from negligence? (b) What is the relationship of medical injury to claims and litigation? (c) What are the costs of medical injury to society? (d) What effect does the threat of litigation have on clinical practice? (Questions (c) and (d) are discussed in later chapters.)

A random sample of 30 195 records of patients admitted to one of 51 hospitals in New York State in 1984 were screened by nurses according to predefined criteria. These criteria included death, increased length of stay, readmission to hospital, fever at discharge, and unplanned transfer to the operating theatre or intensive care unit. 7743 of the records met the screening criteria and were then independently reviewed by two physicians. Preliminary studies had established that reliable and valid judgements could be made in these reviews (Brennan *et al.*, 1989). A study comparing the review method with information available from hospital quality assurance records and risk management and litigation records found that the review process reliably identified the overwhelming majority of adverse events (Brennan *et al.*, 1990).

The study identified 1133 patients (3.7 per cent of all cases) with disabling injuries caused by medical treatment (adverse events). Errors in management were identified in 58 per cent of adverse events. In 27.6 per cent of adverse events the injury was judged to be the result of negligent care. Thus about 1 per cent of admissions resulted in some negligent treatment. In more than 70 per cent of adverse events the injuries led to minimal or moderate disability of less than six months duration, but in 7 per cent it was permanent and 14 per cent of patients died in part as a result of their treatment. Nearly half (48 per cent) of the events were associated with an operation, but drug complications were the most common problem (19 per cent) overall. Standards of care varied considerably, with hospital rates of adverse events varying 10-fold. Adverse events were more common in patients cared for in teaching hospitals but there was less negligence: the higher adverse event rate reflected the type of patient rather than the standard of care.

In the second stage the Harvard team assessed the overall rate of malpractice litigation stemming from these incidents. Using several estimation procedures, based on data from insurance companies and the Department of Health, they calculated that there were 3500–3800 patients claims in the accident year 1984. There were therefore overall about eight times as many adverse events as claims, and probably about 14 adverse events for every paid claim. However, there is a remarkable mismatch between claims and negligent events. They found that only 48 of their sample of injured patients made a claim and only eight of these had suffered a negligent injury. As they detected 313 negligent adverse events this suggests that only about 2 per cent of negligent adverse events result in claims. The remainder of the claims arose out of cases where there was no negligence, though in 10 an injury had occurred. The true odds of a claim following an actual negligent adverse event is much closer to 1:50 than 1:8. The 1:8 ratio arises because many claims are initiated where there is no negligence, though this may not be apparent at the beginning of the legal process. There is certainly, even in the USA, a potential for much more malpractice litigation than is presently experienced. Given the gap between injuries and litigation, it is difficult to argue that tort litigation is adequately compensating individuals or deterring poor medical practice.

The authors concluded that medical injury in hospitalized patients is more frequent than is generally recognized, and that reduction of iatrogenic injury requires both improved methods of detection and the development of mechanisms to prevent errors (Leape *et al.*, 1991). If their results are extrapolated to the UK, with approximately 8 million hospital admissions per annum in England alone (DHSS, 1991), they suggest that about 300 000 adverse events occur in England per annum from hospital admissions alone, with 75 000 being due to negligence. One must remember, however, that 70 per cent of these involve only minor or short-term injury.

Methods of studying medical accidents

The Harvard study is the most comprehensive study of medical accidents yet published. However, many other studies bear on the problem of medical accidents. At one extreme are broad-brush epidemiological analyses and at the other studies of individual cases that provide a picture of the evolution of an accident and the factors that contributed to the final injury. Much information seems to be collected without any clear idea of how it might be analysed, and what implications can be drawn from it. Studies that attempt to understand how medical accidents occur are still rare. Anyone concerned with medical accidents, whether from a management, clinical, or research perspective, needs to consider which method of study will provide the best information for their particular purpose. A variety of methods is needed in research into medical accidents, each having its particular advantages and disadvantages.

Fine-grained analyses of causes, such as those described in the previous chapter, and effective well-targeted risk management, or other interventions aimed at reducing accidents, both depend upon first collecting detailed and reliable accounts of accidents. In some studies the data are obtained from medical records or standard report forms, while in others interviews and direct observation are employed. The following sections review a number of different types of study: occurrence screening, confidential enquiries, audit, studies of errors, and studies of individual accidents and claims. Results of studies are discussed, but mainly to give a broad indication of the method's scope and limitations. Many of the findings will be discussed in detail in later chapters.

Analyses of medical records: occurrence screening

Occurrence screening, like the Harvard study, is a technique in which medical records are reviewed to identify poor-quality care. It is based on two main principles: firstly, that it is far more practical to specify and describe what does not constitute good quality care than to specify what does; and, secondly, that focusing on poor-quality care is an effective way to bring about improvement in overall standards. Adverse occurrences include cases in which a patient experiences an adverse event as defined in the Harvard study, but the screening criteria are much broader. Readmission for incomplete previous management, unplanned return to operating theatre, cardiac or respiratory arrest, transfer to special care unit, and abnormal results not accessed by physician are all included. Adverse occurrences are predefined as being of interest regardless of whether the patient was injured by treatment (Bennet and Walshe, 1990).

These systems were developed in the USA, and have been extensively used as part of overall quality-control management. In the UK a project has recently been completed by Brighton Health Authority. Over 6000 sets of records were screened in three specialties — ophthalmology, obstetrics, and orthopaedics. Preliminary results indicate that, in orthopaedics and obstetrics, over half of the records showed at least one 'occurrence'. Between 10 and 25 per cent of patients had multiple occurrences. Analysing the occurrences in one specialty — ophthalmology — by type shows that the commonest occurrences were also the more trivial ones, particularly medical and nursing record deficiencies. However, clinically relevant occurrences made up 37.2 per cent of the total, with the commonest occurrences being problems related to cataract extraction, certain specific ophthalmic complications, and unplanned removal/injury during repair surgery (Brighton Health Authority, 1992).

Some of the advantages of occurrence screening are that it can be implemented across a range of specialties, or a whole hospital, relatively easily. Because it is, by nature, a multidisciplinary approach, it can identify issues which cross professional boundaries. It provides targets for further, more specific audits, which focus on clinically relevant events. However, in its present form it takes no account of the outcome for the patient and is largely confined to hospital-based services.

Analyses of closed claims

Closed-claims analyses are investigations of cases that have resulted in litigation which, as we have seen, constitute only a small proportion of adverse events. The largest study is that of the American Society of Anesthesiologists, which started in 1985 and has run continuously since then. A standard report form is completed by an anaesthetist reviewer for every claim where there is enough information to reconstruct the sequence of events and determine the nature and cause of the injury. Typically, a closed-claim file consists of the hospital record, the anaesthetic record, narrative statements of the staff involved, expert and peer reviews and reports of both the clinical and legal outcome. An initial report (Cheney *et al.*, 1989) described the most common complications and injuries; in more than half of the cases the anaesthesia-related injuries were severe and disabling or resulted in death. More recent reports have focused on specific situations, in an effort to draw lessons of clinical significance. However, in an analysis of adverse respiratory events Caplan *et al.* (1990) found that the 'distinguishing feature in this group of claims was the reviewer's inability to identify a specific mechanism of injury' — only 9 per cent of these claims involved obviously inadequate behaviour — although there was widespread agreement that better monitoring would have prevented the complication.

Two analyses of expert reviews of British obstetric claims have also been carried out. Cases were drawn from the file of the Medical Protection Society, a professional association providing indemnity cover for doctors (Ennis and Vincent, 1990) and from Action for Victims of Medical Accidents, a patients' charity with access to medical reports (Vincent *et al.*, 1991). The cases considered were those where the outcome was either (a) stillbirth, (b) perinatal or neonatal death, (c) CNS damage leading to handicap, or (d) maternal death. Human error was frequently implicated in these obstetric accidents and many were avoidable. Junior and middle-grade staff were most frequently involved and three major areas of concern were identified: inadequate fetal monitoring, mismanagement of forceps (undue traction, too many attempts), and lack of involvement of senior staff. Problems included an inability to recognize abnormal or equivocal fetal heart traces and a failure to take appropriate action when an abnormal trace was recognized. Inexperienced doctors were frequently left alone for long periods in labour wards, even in cases where consultants had previously expressed doubts about their competence. Moreover, junior doctors frequently had difficulty in obtaining prompt assistance when difficulties arose. Although these incidents represented a tiny fraction of the 3 million safely delivered babies born in the period of these studies, the authors suggested that they revealed more general problems. Further studies on training and fetal heart monitoring by Ennis (1991) supported this viewpoint.

Analyses of malpractice claims have also been used to investigate the relationship between individual physicians' competence and their malpractice experience. Sloan *et al.* (1989), in an analysis of claims in Florida between 1975 and 1988, found little relationship between claims experience and any standard indicators of physicians' competence, such as qualifications or prestige of medical school attended. Physicians with adverse claims experience were slightly more likely to face discipline by the state licensing board, but fewer than 10 per cent were formally disciplined. Female physicians were less likely to be sued than male; Sloan and his colleagues suggest that this is because their style of practice is less likely to result in claims. Overall there seemed little direct relationship between claims experience and competence. Kravitz *et al.* (1991), in a similar study, also found 'little evidence that claims were sufficiently concentrated, either in number or kind, to permit negligence reduction strategies targeted at individuals'. Epidemiological malpractice data might identify a few extremely problematic physicians but its main use is in identifying common and serious errors, and their possible causes. Kravitz *et al.* suggested, for instance, that poor co-ordination between different staff and specialties played a 'strong supporting role' in cases of poor patient management.

Both occurrence screening and claims analyses are fundamentally reliant on the quality of medical records — poor, sketchy, or badly organized records are not a good foundation for any form of records-based audit. In the UK a number of audits of medical records have revealed alarming and persistent deficiencies. In a survey of New Zealand anaesthetists Galletly *et al.* (1991) found that 55 per cent of them occasionally omitted or falsified records. The reasons for omissions were generally that the anaesthetists perceived record keeping as a distraction from patient care; there was no direct evidence of manipulation for medico-legal reasons. In a review of obstetric accidents (Vincent *et al.* 1991) expert reviewers found notes to be illegible in a third of cases, inconsistent in a quarter, and inadequate observations recorded in more than half. There were also suggestions that notes had been tampered with and of a 'highly suspicious selective loss of documents'.

Even with adequate medical records, closed-claims analysis has a number of limitations. Caplan *et al.* (1990) point out that this method cannot provide general estimates of risk, and is also limited by the absence of control groups, a probable bias towards adverse outcomes, and a partial reliance on data from direct participants rather than objective observers. It is true that the analysis of claims can never provide a good estimate of risk, and the other limitations must certainly be acknowledged. Nevertheless, valuable information has been obtained, which has led to more specific and better-controlled studies. A more pertinent question is perhaps whether one needs to wait so long before investigating such incidents. As Caplan *et al.* (1990) point out, the 'objectivity of an eyewitness is degraded by the passage of time, interactions with other observers and premature efforts to reach conclusions'. A better understanding of such events requires investigations that begin immediately after the critical incident or adverse outcome. It is this recognition that partly underlies the development of confidential enquiries into deaths during childbirth and operations, and studies of other kinds of critical incidents.

Confidential enquiries

Confidential enquiries have been conducted into maternal deaths, perinatal deaths, and deaths after surgery. Since the 1950s all deaths in the UK of a mother during labour have been reported. When a death is reported an enquiry form is sent to all health care professionals directly involved in the mother's care. These forms, together with hospital records and inquest reports, are then reviewed by regional obstetric and anaesthetic assessors. The results of these confidential enquiries are published every three years (DHSS, 1989). Maternal mortality has fallen steadily, but all the reports have found that a high proportion of the deaths were avoidable, some being associated with substandard care. For instance, the delegation of high-risk obstetric and surgical operations to relatively junior doctors has been criticized. The reasons for this practice can, however, only be guessed at. The delegation might have been due to staff shortages, over-confidence on the part of the junior or inadequate supervision by the consultant, or a failure of communication between junior and senior doctors.

The confidential enquiry into maternal deaths was a remarkable achievement, and it is taken nearly twenty years for other specialties to follow its lead. Two reports of mortality associated with anaesthesia and surgery (CEPOD) have now been published (Buck *et al.*, 1987; Campling *et al.*, 1992). The first had a mixed reception, some greeting it as an outstanding example of a critical audit, others criticizing its lack of scientific method. Both reports are long, and crammed with information and tables which are not easy to assimilate. Nevertheless, there are some clear findings: 'deficient data from case notes and Hospital Activity Analysis, inadequate provision of emergency services, low necropsy rates, poor supervision of junior staff and the higher risk attached to surgeons operating outside their specialties' (Nixon, 1992). More consultant involvement and more supervision of junior and locum staff are clearly needed. In general the authors are content to present the data and leave others to analyse it more critically, assess avoidability, apportion blame, and identify solutions; the discussion 'avoids contentious issues' (Nixon, 1992).

Critical incident reporting

The studies and surveys reviewed so far have been mainly concerned with a broad description of certain types of adverse events, and with trying to identify factors that are associated with poor-quality care. Useful though such surveys are, both as research and as a quality improvement tool, they tell us little about the genesis of accidents or the full range of factors that may be implicated. An alternative approach is to look in detail at specific incidents in which a patient was harmed, or narrowly avoided harm. The critical incident technique is a common sense method for retrospectively analysing a series of such incidents.

A critical incident is defined as 'an episode of patient care in which the actions of a physician had specific beneficial or detrimental effects on a patient. The term critical simply means that the physician's actions were very likely directly responsible for the effects on the patient' (Anon., 1988). The technique was originally used in medicine for defining core skills for inclusion in a curriculum, but was used by Cooper (1984) in a pioneering study of anaesthetic misadventures, and later in other settings, for example intensive care (Wright *et al.*, 1991).

Reports of accidents and near misses are obtained from those involved, and then analysed for common themes. For instance, Cooper and his colleagues (1984) reviewed 1089 potentially dangerous incidents in anaesthesia, 70 of which had contributed to 'a substantive negative outcome'. The incidents most frequently reported included breathing circuit disconnections, drug-syringe swaps, gas-flow control errors, and losses of gas supply. They classified most errors into failure of technique, errors of judgement, and lapses of attention and made many recommendations for improved patient care including additional technical training, improved supervision, improved organization, and the use of additional monitoring equipment. The authors commented that previous studies of anaesthetic risk looked only at patient characteristics and equipment. They emphasized the need to study the actions of the anaesthetists — the human factor.

Wright *et al.* (1991) studied critical incidents in the intensive therapy unit and found that inexperience with equipment and shortage of trained staff were factors felt most often to contribute to such incidents. The majority of incidents, even in this 'high-tech' specialty, were considered to be due to human error. They point out that this technique offers no information about the frequency of such events, though they suspect they are common. Morbidity and mortality meetings may also analyse critical incidents and areas of weakness, but the approach is less systematic and they can produce a public apportioning of blame that does not help understanding or promote acceptance of responsibility for error.

The critical incident technique has many advantages. It does not require that a patient is actually injured, it monitors all aspects of care, it focuses on the prevention of incidents, and can, to some extent, monitor the effects of any preventative strategy (Williamson, 1988). Some authors have suggested that the technique should be used continuously, but it is not clear how practical this is. The principal problem with critical incident studies is maintaining the interest and enthusiasm of the staff involved, and fostering a climate in which incidents are reliably reported and analysed. Unless staff are actively involved and see such studies as directly relevant to clinical practice, completing critical incident forms is seen as yet another time-consuming chore, with a consequent decline in participation.

Medical audit

Medical audits aim to assess and improve patient care, enhance medical education, and improve efficiency. One would imagine that, with these aims, audits would identify accidents and errors even though this is not their primary purpose. Yet most audits stop short of identifying mistakes, or at least stop short of publishing the information. For instance, in an informal review of 20 clinical audits published in the British Medical Journal, less than half the studies made any assessment of the adequacy of care (Vincent *et al.*, 1992). There are, however, some important exceptions. For instance, Sharples *et al.* (1990) studied avoidable factors contributing to the death of children with head injuries. They found that avoidable factors were present in 32 per cent of cases, identifying failures of diagnosis or delayed recognition of intracranial haemorrhage, inadequate management of the airways, and poor management of transfer between hospitals as key problems. Dearden and Rutherford (1985) carried out a prospective study of the treatment of severe trauma, setting out to identify areas where improvements could be made. Common problems identified were missed or delayed diagnosis, a failure to recognize that treatment was necessary, and an inability to carry out a necessary treatment. As a result of her previous study Dearden (1990) introduced a triage system for severely injured patients, and a requirement that a senior doctor be called if a patient scored above a specified point on the revised trauma scale. In a later study they reported a substantial reduction in errors. Both Sharples and Dearden set out to examine errors and inadequate care. Their results consequently had clear implications for improving clinical practice.

Studies of errors

Studies of errors in medicine take various forms. There is a large and important experimental literature on biases and errors in clinical judgement which is primarily concerned with an understanding of the process of medical thinking and its relation to diagnostic accuracy and clinical competence (see Chapter 8). This section is concerned only with studies of errors in actual practice, their nature and frequency.

Studies of the accuracy of tests and clinical procedures certainly suggest that errors in medicine are common, though accuracy is obviously very much affected by the limitations of particular tests. For instance, we studied the ability of junior doctors working in accident and emergency to detect radiographic abnormalities (Vincent *et al.*, 1988). Senior house officers missed 35 per cent of radiographic abnormalities and 39 per cent of abnormalities with clinically significant consequences. This suggested that a patient presenting at an Accident and Emergency department for X-ray at a time when no radiologist is available stood a one in three chance of an abnormality being missed. High rates of error have also been found in detecting the clinical signs of cyanosis, interpreting electrocardiograms and radiographs, taking a history, and assessing biopsy specimens in the laboratory. Comparisons of diagnoses with necropsy results or of initial diagnoses with final diagnoses also show high levels of inaccuracy (Vincent, 1989). Such studies can in turn be used to examine factors associated with diagnostic inaccuracy. For instance, Zarling *et al.* (1983) found that accuracy was adversely affected by an over-reliance on diagnostic testing. Inaccurate diagnoses and test results do not necessarily imply human error, but they do show that the potential for accidents and adverse events is always present, perhaps to a higher degree than is generally realised.

It is useful to expose errors though only if a climate exists in which errors can be admitted without recrimination. Wu *et al.* (1991) analysed 114 mistakes made by house officers, 90 per cent of which had a serious outcome. They asked about causes of the mistakes — fatigue and too many concurrent tasks were frequently cited — and about reactions to the error. Only half the juniors talked to their immediate senior, but 88 per cent discussed the incident with some senior doctor. Constructive changes in practice were more likely if responsibility for the mistake was accepted. However, accepting responsibility was also associated with emotional distress. In a few instances psychological problems, for example a nervousness about clinical decisions, persisted.

Observational studies

Studies of accidents and adverse events can suggest that the skills of some doctors may be deficient, which in turn leads to more specific studies. Sometimes, as in the study of X-rays cited above, these can be carried out from medical records. For some skills however, direct observation is necessary. For instance, Skinner (1985) observed preregistration house officers carrying out cardiopulmonary resuscitation; their skills fell well short of the expected standard. Marteau *et al.* (1987) found that experience of resuscitation increased confidence, but not skill, in nurses. The advantage of such observational studies is that one can examine the process of clinical care and the ways in which errors are made and corrected. A fascinating extension to this kind of work is the use of simulated patients (a mannequin and all the usual monitors, together with simulated output) to study the errors made in anaesthetic process, and the critical incidents that they lead to. These methods are useful in studying the 'chain of accident evolution' and the techniques subjects use to avert or recover from problems and incidents (DeAnda and Gaba, 1990, 1991). Video recording, 'think aloud' techniques, and careful debriefing greatly increase both the quality and quantity of information collected. Similar methods are common in aviation and the nuclear industry, but have not yet been widely applied in medicine.

There have also been some attempts to directly monitor errors in patient care. As accidents are comparatively rare, an observational study would seem to be an unrewarding prospect. One might have to wait quite a long time before observing even a minor error in the care of a patient. Potentially, however, observation of errors allows a much more accurate estimate of their true incidence.

No study to my knowledge has tried to directly observe patient care. However, Stocking and her colleagues (1992) used observers in routine meetings to record 'eyebrow raising events' in which a member of staff identified a situation in which a patient's care was inappropriate. The opportunity for the study arose when a senior surgeon was admitted to his own hospital and observed numerous errors in his own care! The full results are not yet available but about 1.3 such events were recorded for each patient admission. Half of these events were later judged to be errors. About 20 per cent were serious resulting in at least temporary disability.

There is one other form of error observation that does not seem to have been tried, and which could have considerable potential — observation by the patient. Clearly, patients are unlikely to be able to judge the adequacy of diagnoses or clinical skills, but there are many areas where they could provide valuable information. Anecdotal reports suggest that many patients are very much aware of mistakes made in their treatment, and that they often have to point them out. They may realise, for instance, that an X-ray is inappropriate during pregnancy, that an angiogram is inappropriate for a fractured pelvis (taken to the wrong X-ray department), that they have been given the wrong medication, or that there has been no communication between doctors on different shifts. Clearly these observations could be compared with staff perceptions and some validation obtained.

Why study errors and accidents?

There are various arguments against studying medical accidents, or at least against focusing enquiries on such incidents. An issue that is sometimes raised is that it is unwise to publish data on accidents for fear of promoting or facilitating litigation. There are two aspects to this: firstly, that an individual case may be revealed and the patient may sue, and secondly that the more the public realises how often errors occur the more litigation there will be (this latter issue is considered below). Obviously one should never identify an individual case, or provide sufficient information to identify that patient. If an audit does report details of accidents, then they need to be disguised. It is unlikely that research material would be demanded by the courts, but one can protect oneself by making the material anonymous at an early stage so that it can no longer be linked to any particular patient.

A more persuasive argument is that with audit of routine care well established (Moss, 1992) there is no need to specifically examine errors and accidents. Focusing on accidents, it could be said, betrays an unduly negative and critical attitude and is likely to antagonize clinicians and reduce the confidence of patients. I suggest that studying errors and accidents is an essential part of any comprehensive quality initiative and that it can reveal a great deal about ordinary clinical practice. The arguments fall into several broad categories.

(a) *Understanding and prevention*: The principal reason for studying accidents and their causes is that it may help prevent them. Errors may not always lead to accidents but will often cause unnecessary distress, delay to patients and staff, wasting of time, additional expense, and a whole range of minor inconveniences that adversely affect health care. Any systematic attempt to reduce error should have widespread consequences. While it is possible that monitoring routine care in general audits may reduce errors, it is by no means certain. The absence of any attempt to even monitor quality of care in many audits, let alone monitor adverse patient outcomes, suggests that errors and accidents may go unnoticed and unremarked if studies do not specifically set out to investigate them.

(b) *Understanding accidents requires 'backward reasoning'*. Another reason a routine audit is unlikely to reveal the true likelihood of an accident relates to the very nature of accidents themselves, and the way people think about them. Given a description of an accident it is possible to look back and identify the sequence of events and factors contributing to the accident. It is much harder to extrapolate forward and envisage the potential consequences of 'unsafe acts', or even to realize that they are unsafe. Similarly, many of the circumstances of an accident, which include design faults, prior management decisions, and personal characteristics of those involved may be unknown, or at least it will not be appreciated that they could play a role in the evolution of an accident. In addition, some accidents may only occur when an unusual set of circumstances conspire to produce a lethal outcome of an error that normally goes 'unpunished'. A routine audit could reveal such errors, but their potential danger is unlikely to be appreciated. These issues are discussed in much greater depth in Chapter 1.

(c) *An effective way of identifying wider problems*. The study of any aspect of clinical practice could reveal the multitude of influences that determine the final outcome of any particular procedure. The advantage of studying critical incidents or adverse events is that the event in question is obviously important. This in turn means firstly, that it is memorable and so more easily recalled by those involved and, secondly, that the processes it reveals are likely to be important factors determining the quality of care. The study of a relatively small number of accidents or near misses can never provide conclusive proof of more widespread problems, but can at the very least suggest the direction that future investigations should take. Our suspicions that junior doctors received little training in fetal heart monitoring (Ennis and Vincent, 1990) was certainly confirmed by Ennis (1991) in her study of training. Future audits and assessments of skills can be more effectively targeted by studying accidents and critical incidents.

(d) *'Every error is a treasure'*. Studying and being aware of errors in clinical practice, as McIntyre and Popper point out (1983), facilitates learning. Psychologists would accord awareness of errors a similar importance, but characterize it as a form of feedback. Unless one has feedback on one's own performance (in any task) there is little hope of improving performance. A further benefit is that awareness of errors provide a stronger motivation for change than simply a general desire to improve standards. In the audit process, the most difficult part is the intervention that follows the initial audit; here the clinician investigator is faced with changing the behaviour of staff in the unit or the organization, whether this involves further training, change in procedures, or requiring stricter adherence to existing procedures. Frequently, the simple feedback of results does not achieve the desired change, and more specific measures are needed. The study of Wu *et al.* (1991) showed that accepting responsibility for an error was strongly associated with subsequent changes in clinical practice. Making and acknowledging mistakes is one of the most powerful agents for change.

(e) *The effect on staff and patients*. The cost of medical accidents, in both human and financial terms, is huge for both patients and doctors. One of the primary aims and motivations for the Harvard study was to examine alternatives to the tort system of compensation. This required an examination of the incidence of negligent adverse events. If for no other reason adverse events, and their consequences for both patients and doctors, should be studied so that adequate care and compensation for patients, and support for staff, can be planned and made available after accidents have occurred.

(f) *For individual patients and for public confidence*. Many patients who complain or take legal action are not only interested in compensation (see Chapter 14): they are also concerned that the same thing should not happen to anyone else. They want an investigation and for their experience to be drawn to the attention of those involved, in the hope that changes in clinical practice may result.

There is also the wider question of the maintenance of trust and confidence between patients and doctors. Medical accidents must not only be investigated, but be seen to be investigated. Public confidence in air travel is increased by the extensive investigations of accidents, and the apparent determination of airlines to learn from their mistakes. Yet in medicine there is little public acknowledgement of mistakes. While this could be defended on the grounds that it is vital that patients continue to have confidence in their doctors, the actual consequence may be the exact opposite. Patients who are the victims of medical accidents often experience great difficulty in finding out what went wrong. This leads to a loss of trust, a loss of confidence, and sometimes to litigation. It may be reassuring to look to the USA. With the advent of serious quality assurance programmes, and special services for patients who may have sustained an injury, litigation does not seem to have increased (Rosenthal, 1987). Studying medical accidents should ultimately improve the relationship between patients and their doctors.

References

Anon. (1988). Critical questions, critical incidents, critical answers. *The Lancet* (June 18), 1373–74.

Bennett, J. and Walshe, K. (1990). Occurrence screening as a method of audit. *British Medical Journal*, 300, 1248–51.

Brennan, T. A., Localio, A. R., and Laird, N. (1989). Reliability and validity of judgements concerning adverse events suffered by hospitalized patients. *Medical Care* 22, 1148–58.

Brennan, T. A., Localio, A. R., Leape, L. L., Laird, N., Peterson, L., Hiatt, H., and Barnes, B. (1990). Identification of adverse events occurring during hospitalization. *Annals of Internal Medicine* 112, 221–26.

Brighton Health Authority (1992). *Medical audit and occurrence screening.* Preliminary report of CASPE project.

Buck, N., Devlin, A. B., and Lunn, J. N. (1987). *Report of the confidential enquiry into perioperative deaths.* Nuffield Provincial Hospitals Trust/King's Fund, London.

Campling, E. A., Devlin, H. B., Hoile, R. W., and Lunn, J. N. (1992). *Report of the national confidential enquiry into perioperative deaths 1990.* National Confidential Enquiry into Perioperative Deaths, London

Caplan, R. A., Posner, K. L., Ward, R. J., and Cheney, F. W. (1990). Adverse respiratory events in anaesthesia: a closed claims analysis. *Anesthesiology* 72, 828–33.

Cheney, P. W., Posner, K., Caplan, R. A., and Ward, R. J. (1989). Standard of care and anaesthesia liability. *Journal of the American Medical Association* 261(11), 1599–1603.

Cooper, J. B., Newbower, R. S., and Kitz, R. J. (1984). An analysis of major errors and equipment failures in anaesthesia management: considerations for prevention and detection. *Anesthesiology* 60, 34–42.

DeAnda, A. and Gaba, D. M. (1990). Unplanned incidents during comprehensive anaesthesia simulation. *Anesthesia and Analgesia* 71, 77–82.

DeAnda, A. and Gaba, D. M. (1991). Role of experience in the response to simulated critical incidents. *Anesthesia and Analgesia* 72, 308–15.

Dearden, C. H. and Fisher, R. B. (1990). Improving the care of patients with major trauma in the accident and emergency department. *British Medical Journal* 300, 1560–63.

Dearden, C. H. and Rutherford, W. H. (1985). The resuscitation of the severely injured in the accident and emergency department — a medical audit. *Injury* **16**, (249–52).

Delamothe, T. (1989). Defence wars. *British Medical Journal* **298**, 699–700.

DHSS (1989). *Report on confidential enquiries into maternal deaths in England and Wales*. HMSO, London.

DHSS (1991). *Hospital inpatients enquiry summary tables*. HMSO, London.

Ennis, M. (1991). Training and supervision of obstetric senior house officers. *British Medical Journal* **303**, 1442–43.

Ennis, M. and Vincent, C. A. (1990). Obstetric accidents: a review of 64 cases. *British Medical Journal* **300**, 1365–67.

Galletly, D. C., Rowe, W. L., and Henderson, R. S. (1991). The anaesthetic record: a confidential survey on data omission or modification. *Anaesthesia and Intensive Care* **19**(1), 74–78.

HMPS (1990). *Patients, doctors and lawyers: medical injury, malpractice litigation and patient compensation in New York*. Report of the Harvard Medical Practice Study.

Kravitz, R. L., Rolph, J. E., and McGuigan, K. (1991). Malpractice claims data as a quality improvement tool. *Journal of American Medical Association* **266**(15), 2087–92.

Leape, L. L., Brennan, T. A., Laird, N., Lawthers, A. G., and Hiatt, H. (1991). Adverse events and negligence in hospitalized patients. *Iatrogenics* **1**, 17–21.

McIntyre, N. and Popper, K. (1983). The critical attitude in medicine: the need for a new ethic. *British Medical Journal*, **287** 1919–23.

Marteau, T. M., Wynne, G., Johnston, M., Whiteley, C. A., and Evans, T. R. (1987). Inability of trained nurses to perform basic life support. *British Medical Journal* **294**, 1198–99.

Mills, D. H. (1978). Medical insurance feasibility study. *Western Journal of Medicine* **128**, 360–65.

Moss, F. (1992). Quality in health care. *Quality in health care* **1**, 1–3.

Nixon, S. J. (1992). NCEPOD: revisiting perioperative mortality. *British Medical Journal* **304**, 1128–29.

Rosenthal, M. M. (1987). *Dealing with medical malpractice: the British and Swedish experience*. Tavistock Publications, London.

Sharples, P. M., Storey, A., Aynsley-Green, A., and Eyre, J. A. (1990). Avoidable factors contributing to death of children with head injury. *British Medical Journal* **300**, 87–91.

Skinner, D. V. (1985). Cardiopulmonary skills of preregistration house officers. *British Medical Journal* **290**, 1549–50.

Sloan, F. A., Mergenhagen, P. M., Burfield, W. B., Bovbjerg, R. R., and Hassan, M. (1989). Medical malpractice experience of physicians. *Journal of the American Medical Association* **262**, 3291–97.

Stocking, C. B. (1992). Echoes of error — reflections of response. Paper presented at *Oxford-Ohio conference*. Wolfson College, Oxford.

Vincent, C. A. (1989). Research into medical accidents: a case of negligence? *British Medical Journal* **299**, 1150–53.

Vincent, C. A. (1992). Medical accidents. In *Medical audit and accountability* (ed. R. D. Mann and A. J. Vallace-Jones), pp. 219–28. Royal Society of Medicine Services International Congress and Symposium Series No.190. Royal Society of Medicine Services Ltd, London.

Vincent, C. A., Driscoll, P. A., Audley, R. J., and Grant, D. S. (1988). Accuracy of detection of radiographic abnormalities by junior doctors. *Archives of Emergency Medicine* 5, 101–109.

Vincent, C. A., Martin, T., and Ennis, M. (1991). Obstetric accidents: the patient's perspective. *British Journal of Obstetrics and Gynaecology* 98, 390–95.

Williamson, J. (1988). Critical incident reporting in anaesthesia. *Anaesthesia and Intensive Care* 16, 101–103.

Wright, D., Mackenzie, S. J., Buchan, I., Cairns, C. S, and Price, L. E. (1991). Critical incidents in the intensive therapy unit. *The Lancet* 338, 676–78.

Wu, A. W., Folkman, S., McPhee, J., and Lo, B. (1991). Do house officers learn from their mistakes? *Journal of the American Medical Association* 265(16), 2089–94.

Zarling, E. J., Sextan, H., and Mihar, P. (1983). Failure to diagnose acute myocardial infarction: the clinico-pathologic experience at a large community hospital. *Journal of American Medical Association* 250, 1177–81.

3

Errors and accidents in obstetrics

JAMES O. DRIFE

Historical background

Childbirth in developed countries is now very safe. Every year in Britain there are more than 600 000 births, most of them in NHS hospitals. Of the mothers, only about fifty die as a result of pregnancy (Department of Health, 1989). Of the babies, over 99 per cent are born alive, and over 97 per cent without major handicap. These statistics contrast sharply with those of only sixty years ago.

Maternal mortality. This is 'the death of a woman while pregnant or within 42 days of termination of pregnancy, irrespective of the duration and the site of the pregnancy' (Department of Health, 1989). In the early 1930s in Britain, maternal mortality occurred once in every 200 births — as it had done since records began. Today in parts of Africa and in the favellas of Rio de Janeiro, the maternal mortality rate is still 1 in 200 (Laguardia *et al.*, 1990; WHO, 1991). In Britain the rate began to fall rapidly after 1937, and nowadays it is less than 1 in 10 000. In a large British maternity unit with 5000 births a year, one maternal death can be expected every three years on average. Sixty years ago, it would have been one a fortnight.

Perinatal mortality. The perinatal mortality rate — the number of stillbirths and deaths during the first week of life — is now under 10 per thousand births in Britain. In 1930 it was over 60 per thousand. Some European countries have lower perinatal mortality rates than Britain's, but some are higher: Greece and Portugal, for example, have perinatal mortality rates around 15 per thousand (Chamberlain, 1991b).

Causes of the improvement. The reasons for the increased safety of childbirth are many, and include a reduction in family size and better general health and living standards. Social factors alone, however, do not explain the improvement. For example, living conditions in Britain did not improve greatly during World War II but the maternal mortality rate fell sharply between 1937 and 1945. Medical developments such as antibiotics, blood transfusion, and safe anaesthesia made substantial contributions, as did the increasing expertise of midwives and obstetricians. The Central Midwives Board was set up in 1902 and a salaried midwifery service was proposed in 1935 (Grant,

1981). The British (now Royal) College of Obstetricians and Gynaecologists was founded in 1929.

Specific professional and political initiatives also helped. In 1952 the Confidential Enquiry in Maternal Deaths was started, to analyse the reasons for maternal deaths and recommend improvements in practice. In 1970, due to reductions in other causes of death, criminal abortion was the leading cause of maternal death in Britain: in 1967 the Abortion Act came into force, and by 1982 criminal abortion had been eliminated as a cause of maternal death (Department of Health, 1989).

Attitudes. The dramatic improvement in safety of pregnancy has led to a change in people's expectations. It is taken for granted that a woman will survive childbirth, and it is expected that the baby will be healthy. The high risks of sixty years ago are forgotten, and obstetricians increasingly have to justify their interventions. It is ironic that British obstetricians are sometimes advised to learn from birth practices in developing countries where rates of maternal and perinatal mortality are similar to those in Britain a century ago.

Increasing litigation

High expectations create the feeling that if pregnancy has an adverse outcome someone must be to blame. Obstetricians have helped to foster this climate of opinion. Because their goal is to reduce the risks of pregnancy to zero, they feel a sense of failure over an adverse outcome and this feeling may communicate itself to their clients. When maternal or fetal death occurs, obstetricians try to identify avoidable factors and it is not surprising that clients should do the same (Ennis *et al.*, 1991).

Most obstetricians are also gynaecologists. Gynaecological practice, like obstetrics, often involves healthy women and has to live up to high expectations. For example, over 90 000 female sterilization operations are now performed every year in Britain (Drife, 1988a). Patients expect this operation to have a perfect outcome, and if a failure occurs the patient's first reaction is often to blame the doctor (Argent, 1988).

Number and size of claims. Before the advent of NHS Indemnity, obstetrical and gynaecological claims constituted around 20 per cent of the workload of the Medical Protection Society: of these, 40 per cent were obstetrical and the remainder gynaecological (Brown, 1985). The Medical Defence Union in the mid-1980s was opening 9000 new litigation files per year, and some 600 of these related to obstetrics and gynaecology (Symonds, 1985).

Obstetric cases may involve very large settlements. A child who requires constant nursing and who has a normal expectation of life may be awarded over a million pounds in damages. These large sums are newsworthy, and publicity leads to other parents turning to litigation. The cost of obstetric cases was a major factor in the rapid increase in the cost of subscriptions to medical defence organizations before the introduction of NHS Indemnity in January 1990. These settlements damage the public image of obstetrics: a case may be settled because of a minor or indeed debatable lapse in clinical care, but a large award gives the impression of serious clinical incompetence.

Recruitment. In the USA the increase in litigation against obstetricians has led to some specialists abandoning obstetrics and concentrating on gynaecology. Obstetricians and gynaecologists, 5.2 per cent of all US doctors in 1984, accounted for 26.9 per cent of total indemnity payments, and claims against obstetricians and gynaecologists increased by over 50 per cent between 1981 and 1985 (Quam *et al.*, 1988). Recruitment of medical graduates into obstetrics and gynaecology in Britain is now a cause for concern in the Royal College of Obstetricians and Gynaecologists, and the threat of litigation is often cited as a major reason. However, a survey by the College showed that although 85 per cent of British obstetricians are or have been involved in litigation, this is less of a deterrent to recruitment than the long working hours, resident conditions, and job prospects of junior hospital doctors (Saunders, 1992).

Structure of pregnancy care in Britain

International comparisons. Patterns of obstetric care vary considerably between different countries, even in the developed world. In the Netherlands 35 per cent of births are in the woman's home (Treffers and Lann, 1986) but in Denmark 99 per cent are in hospital (Scherjon, 1986). In some countries much of the care is given by midwives, but in others medical intervention is more common. In the USA rates of obstetric intervention are correlated with the social class of the patient, and there are higher rates of Caesarean section among those insured for private health care (de Regt *et al.*, 1986). Among middle-class women in Brazil who attend private obstetricians, Caesarean section rates as high as 92 per cent have been reported (WHO, 1991).

Antenatal care. In Britain there is more uniformity of obstetric care throughout the country. The pattern of antenatal clinic visits — monthly in the first two-thirds of pregnancy and increasingly frequent thereafter — involves a sharing of responsibility between hospital, general practitioner, and community midwife. This traditional pattern is not based on scientific evidence (Chamberlain, 1991a), and new patterns are now being introduced, involving greater autonomy for midwives.

Care in labour. Home delivery accounts for less than 1 per cent of deliveries in Britain, and private obstetric practice for a small number. The great majority of British babies are born in NHS hospitals, and around 70 per cent are delivered by midwives — usually a hospital midwife. Flexible schemes are now being introduced, such as the so-called 'domino' scheme, in which the community midwife accompanies the woman to hospital and supervises the delivery and the return home of mother and baby a few hours later. Such systems still account for only a minority of deliveries (Smith and Jewell, 1991), partly because it is difficult to organize flexible teams of personal midwives for the 600 000 women who deliver in Britain every year. Most women are delivered by a hospital midwife whom the woman has never met before. Hospital midwives work eight- to twelve-hour shifts which may change during the woman's labour. Maternity hospitals are staffed to the average, rather than the maximum, workload, so at busy times a midwife may have to look after more than one woman in labour.

Junior doctors. If a problem arises the midwife is required to call a doctor. The medical staffing structure of a British hospital is hierarchical and the most junior doctor is first on call, often supported by a middle-grade doctor resident in the hospital, with the consultant being on call from home. The junior doctor will have been qualified for at least one year and will usually be training for a career in obstetrics or general practice. Junior posts last six months, so the doctor called by the midwife often has only a few months of obstetric experience.

Middle-grade doctors. A middle-grade doctor — a registrar or senior house officer — will have had at least one year's and often several years' experience of obstetrics. There is a limit on the number of registrar posts in British hospitals, because the number is linked to the number of consultant vacancies. Some registrars are doctors from overseas, gaining experience in Britain before returning to their own countries. At present, overseas doctors come mainly from the old commonwealth, though increasing numbers are expected to come from Europe.

The registrar may be studying for the Membership examination of the Royal College of Obstetricians and Gynaecologists, which requires at least two years' experience in the specialty after registration, plus another year of experience in a related specialty or in research. He or she may have duties in more than one hospital simultaneously: for example, gynaecological emergencies may be admitted to one hospital and obstetric cases to another several miles away. The Royal College of Obstetricians has strongly urged that registrars should not have duties on more than one site simultaneously.

Consultants. Consultant involvement in the day-to-day running of the labour ward is variable. Antenatal clinics nowadays usually have a consultant in attendance in addition to the junior staff, but the labour ward may be run entirely by junior doctors and midwives, with the consultant attending only when called. The Royal College of Obstetricians and Gynaecologists has recommended that all labour wards should now have dedicated consultant sessions, that is, that the consultant should not have duties elsewhere when on call during the day for the labour ward, but this recommendation has been implemented in few hospitals as yet. In some hospitals the consultant carries out a routine ward round in the delivery suite every morning, but there is no official requirement for such a system.

Relationships between professionals. The relationship between doctors and midwives can be difficult (McKee *et al.*, 1992). Midwives are independent practitioners but they are bound by the Midwives' Rules to call a doctor when they judge this necessary (Drife, 1988b). It is an anomaly that an experienced midwife has to call an inexperienced doctor for advice. Usually the system works because doctor and midwife recognize the reality of the situation and an inexperienced doctor will accept guidance from an experienced midwife. However, if midwives call doctors 'just to cover themselves' the doctor may feel he or she has been bothered unnecessarily. Conversely, a midwife who believes intervention is needed may be frustrated by a doctor who does not take the problem seriously enough. Because of the hierarchical system, a midwife may be reluctant to 'go over the head' of a junior doctor to a more senior member of the medical staff.

Attempts to remedy this problem by regular consultant ward rounds may not be welcomed by the midwives. Many midwives feel that their role as independent practitioners has been eroded by the medicalization of childbirth, and prefer to call a doctor if they need to, rather than having their work regularly checked. Striking a balance between the midwives' wishes and the optimal requirement of safety should not be a difficult task, but there is sometimes a lack of sensitivity by each profession to the needs of the other (Drife, 1989b). This problem has not been helped by poor contact in the past between the respective Royal Colleges of obstetricians and midwives, though a constructive relationship is now beginning to develop.

Definition of 'error' in obstetrics

Definition of errors and accidents in obstetrics is difficult. Defining them as stillbirth or handicap implies that all such outcomes are preventable. This assumption spurs obstetricians to try to improve their practice, but not every accident of nature is a preventable error. For example, most cases of congenital impairment of cerebral function (misleadingly called 'brain damage') are due to unknown or genetic causes, and not to oxygen deprivation during labour (Illingworth, 1985). No more than 10 per cent of cases of cerebral palsy are due to intrapartum asphyxia (Lamb and Lang, 1992).

If there were a universally agreed standard of obstetric practice, lapses could be identified and defined as errors. However, there is little consensus about obstetric management. There is disagreement about which investigations are carried out at the antenatal clinic (Heringa and Huisjes, 1988), and about what constitutes abnormal labour. Obstetric case notes vary greatly between different hospitals, and information considered essential in one hospital may not be recorded in another.

Intervention rates. Variation in obstetric practice between hospitals is reflected by differences in the rates of Caesarean section and instrumental delivery (Paterson *et al.*, 1991). For example, in Dublin in 1983 the rate of forceps delivery was 6.5 per cent in one hospital and 16.3 per cent in a neighbouring hospital (which also had a much higher rate of epidural anaesthesia). In the same year at Birmingham Maternity Hospital the forceps rate was 21 per cent. The perinatal mortality rates in all three hospitals were similar (Drife, 1985). Part of the difference may be due to the fact that different hospitals serve different populations, some with a high proportion from ethnic minority groups, but this is unlikely to explain such wide variations in clinical practice.

Rates of forceps delivery fell sharply in the late 1970s in British hospitals, but rates of Caesarean section have risen steadily from 6–7 per cent in the mid-1970s to 10–15 per cent in the early 1990s (Derom *et al.*, 1988). In the USA in 1980 the Caesarean section rate was 18 per cent: rising rates of Caesarean section coincided with a fall in perinatal mortality, and it was claimed that the increased intervention was responsible for the improved safety of labour. However, O'Driscoll and Foley (1983) pointed out that the National Maternity Hospital in Dublin experienced a fall in perinatal mortality similar to that in the USA, with no increase in its Caesarean section rate.

Protocols. Many British maternity hospitals now have guidelines, varying from general outlines of how normal labour should be managed to detailed instructions for the management of normal and abnormal cases. National bodies are reluctant to endorse such guidelines, partly because national consensus would be difficult to achieve and partly because a national guideline would make it easier for plaintiffs to criticize management that differed from that prescribed. Another reason for reluctance to publish national 'rules' is that they might militate against progress: innovations in practice would be more difficult to introduce if they went against formal guidelines. Some doctors and midwives dislike guidelines because they may limit the extent to which a midwife can use her own judgement, though it is possible to provide guidelines that maximize freedom of choice while at the same time setting out limits of safe practice.

An underlying tension in discussions about the management of labour is the debate between 'natural childbirth' and 'active management of labour'.

Advocates of active management claim convincingly that it reduces the need for Caesarean section (O'Driscoll and Meagher 1980) but women's groups claim that the policy involves too much intervention. For example, O'Driscoll teaches that if the rate of cervical dilatation is less than 1 cm an hour, labour should be accelerated by intravenous oxytocin. Many women, however, are unhappy about having a 'drip' and would prefer a less rigorous definition of slow labour.

Effective Care in Pregnancy and Childbirth was the title of a book published in 1989, which applied scientific analysis to the debates about different styles of practice. The authors attempted to review all published and indeed unpublished trials of obstetric management (Chalmers *et al.*, 1989). The project has a continuously updated database in electronic form, and the book includes lists of interventions that are of proven effectiveness, those that are unproven, and those that have been proved to be ineffective. It represents the most comprehensive attempt yet to assess obstetric interventions in an objective way (Chalmer and Chalmers,1989) and it has been called 'the most important book in obstetrics to appear this century' (Paintin, 1990). Nevertheless, it is extremely difficult to remove bias completely, either from individual trials or from the process of reviewing them and drawing clinically useful conclusions.

Audit in obstetrics

Audit is being introduced into all clinical specialties. The results of clinical management are assessed, weaknesses identified, changes planned and implemented, and the results of the new management assessed. Obstetric audit has been carried out for many years using various strategies.

Confidential enquiries into maternal deaths. In Britain information is collected about every maternal death. A form is filled in and details identifying the woman, the hospital, and the doctors are removed, so that doctors feel free to give as complete an account as possible of the events involved in the death. The forms are scrutinized by Regional and Central Assessors, and every three years a report is produced which points out clinical and administrative lessons and identifies 'avoidable factors'. As numbers of maternal deaths fall, it becomes harder to ensure anonymity, particularly if there were unusual circumstances, and it is now feared that the assessors' comments could be used by lawyers. This would jeopardise the enquiry, as doctors would be reluctant to give information. The enquiry has been most effective in increasing the safety of childbirth and it would be a pity if it stopped.

Perinatal mortality meetings. Most maternity hospitals hold regular meetings at which doctors and midwives review cases of stillbirth or neonatal death. The purpose is to identify practices which could be altered to prevent similar events in the future (Kirkup, 1990). It is sometimes difficult to avoid excessive self-criticism in these meetings, but they are an essential part of audit in large maternity hospitals.

'Near misses'. As obstetric care has improved, traditional perinatal mortality meetings have become less useful for audit. Many of today's stillbirths are babies of extremely low birthweight who might have been classified as abortions only a few years ago, and others involve congenital abnormalities incompatible with independent life. It has therefore been suggested that instead of — or in addition to — the perinatal mortality meeting there should also be a meeting to discuss cases in which the baby has been delivered in poor condition or in which the staff have perceived some problem (Barron, 1991). As yet, not all hospitals hold such meetings.

Fetal monitoring

The obstetrician has two patients and the less accessible is at much higher risk. Methods of assessing the condition of the fetus are still far from perfect.

Antenatal monitoring. During pregnancy, fetal well-being is checked by clinical assessment of the uterus, which is unreliable. Abdominal palpation will miss about 25 per cent of cases of growth retardation as defined by the baby's birth weight in relation to a normal population. Biochemical tests of placental function measure placental hormones in maternal urine or blood. It is now recognized that such tests are inaccurate, because the hormone-secreting function of the placenta may not be related to its ability to transfer nutrients to the baby.

Nowadays the condition of the baby during pregnancy is checked by cardiotocography, by ultrasound measurements of the fetal head, abdomen and legs, and sometimes by ultrasound assessment of blood flow in the umbilical or other vessels. These investigations require the mother to spend time in the ultrasound department or fetal assessment unit, and are therefore not used in the routine antenatal care of all pregnant women but only for high-risk cases.

Monitoring during labour. The fetal condition in labour is assessed by means of the fetal heart rate. This is auscultated by the midwife every fifteen minutes ideally, though because many women walk around during labour, regular checks can be difficult to achieve, especially in early labour. The other traditional way of checking the fetus is to observe the colour of the amniotic fluid, which is normally clear but turns green if the fetus releases meconium (bowel contents) into the fluid. This may be a sign of lack of oxygen. This is one of the reasons for recommending artificial rupture of the membranes early in labour, but nowadays some women are unhappy to have this done as they feel it makes labour unnatural.

Fetal distress. Abnormalities of the fetal heart rate and meconium staining of the amniotic fluid are traditional signs of 'fetal distress'. This is an unsatisfactory term which lacks a single precise definition. It could be defined as lack of oxygen, and the oxygen concentration in the fetal blood can be measured by fetal scalp blood sampling during labour. Indeed, fetal blood can be sampled before labour by ultrasound-guided insertion of a needle into the uterus, but this technique carries risks and is available in only a few specialist centres. More commonly, the term 'fetal distress' means abnormalities which are suggestive but not diagnostic of oxygen lack. It is also used to describe abnormalities of the heart rate detected by electronic fetal monitoring.

Electronic fetal monitoring. For the last twenty years the standard method of assessing fetal well-being in labour has been electronic fetal monitoring (Beard *et al.*, 1971). A continuous recording of the fetal heart rate on a long strip of paper allows assessment of patterns which are related to the fetal condition. A normal pattern involves a rate between 120 and 160 beats per minute, a 'reactive' baseline (that is, good variability of the fetal heart rate from beat to beat), and a lack of decelerations. Decelerations, or 'dips' are a possible sign of fetal compromise, but only if they occur after uterine contractions. Decelerations which are synchronous with contractions ('Type 1 dips') are caused by pressure on the fetal head, and are benign provided that they return to normal as soon as the contraction stops. If there is a delay between the peak of the contraction and the nadir of the deceleration ('Type 2 dips'), there may be fetal compromise. Even a short delay is important, and decelerations may be significant even if the heart rate remains within normal limits during the deceleration.

The most important problem with electronic fetal heart-rate monitoring is that it is not diagnostic of fetal distress. Even with the most sinister combination of abnormalities — a rapid baseline and type 2 dips — there is only a 50 per cent chance that the fetal is suffering oxygen deprivation. Thus if the monitor is used as the sole guide to fetal condition, many unnecessary Caesarean section or instrumental deliveries will be done. Electronic fetal monitoring was always intended to be a screening test to decide which babies should be assessed by fetal scalp blood sampling.

Scalp blood sampling is technically more difficult than electronic fetal heart rate monitoring, as it involves inserting a tubular instrument through the cervix, stabbing the baby's scalp with a small blade and collecting a blood sample in a capillary tube. The blood is then analysed in a machine: a pH of less than 7.20 indicates immediate delivery of the baby. A pH above 7.25 is acceptable, and between those two figures the sample should be repeated within 30 minutes. Because of the technical difficulties of obtaining a sample and maintaining the apparatus, not every unit that uses fetal heart-rate monitoring has access to fetal blood sampling.

Electronic monitoring is important in high-risk cases, but there has been controversy about its routine use in low-risk cases (Friedman, 1986). Most obstetricians have experience of unexpected fetal death in an apparently normal labour, and for this reason it seemed sensible to offer all women the added safeguard of electronic monitoring during labour. The disadvantage is that by detecting abnormalities the chances of unnecessary intervention are increased (Leveno *et al.*, 1986). A large study of electronic fetal heart-rate monitoring in low-risk cases was carried out in Dublin, and showed that the stillbirth rate in the monitored group was no different from that in the unmonitored group (MacDonald *et al.*, 1985).

Nevertheless, it is increasingly common nowadays for a low-risk woman to have electronic monitoring for a short time soon after admission to the labour ward. A normal trace is a considerable reassurance to the woman and her attendants that further monitoring by intermittent auscultation will be all that is required. Newer and more accurate ways of assessing the fetal condition are being investigated (Johnson *et al.*, 1991) but electronic fetal monitoring is likely to be the mainstay of intrapartum assessment for many years yet.

Assessment at delivery. The condition of a baby at delivery is usually assessed by the Apgar score — a simple system which scores the baby's colour, tone, breathing, heart rate, and response to stimulation. The score has little prognostic value, however. Taking a sample of blood from the umbilical cord and measuring its oxygen tension and pH give a more accurate assessment of the baby's metabolic state, but this is far from being routine practice.

One of the best guides to prognosis is the condition of the baby in the first days after delivery. Abnormal neurological signs may amount to 'hypoxic-ischaemic encephalopathy', a condition characterized by fitting, excessive muscular tone, and poor feeding ability (Hall, 1989; Hull and Dodd, 1991). Ultrasound assessment of the brain may show signs of bleeding and later may show cavitation due to lack of oxygen supply. Whether such oxygen deprivation occurred during labour or before labour began can be difficult to tell. Freeman and Nelson (1988) suggested that for 'brain damage' to be attributed to asphyxia four questions should be answered positively:

1. Is there evidence of marked and prolonged intrapartum asphyxia?

2. Did the infant show signs of moderate or severe hypoxic-ischaemic encephalopathy during the newborn period, with evidence also of asphyxial injury to other organ systems?

3. Is the child's neurologic condition one that intrapartum asphyxia could explain?

4. Has the work-up been sufficient to rule out other conditions?

Causes of stillbirth and handicap

The causes of stillbirth and neonatal death include congenital abnormalities, prematurity, antepartum anoxia, and birth injury. At present premature delivery is almost impossible to predict and prevent, but to some extent the other causes are theoretically preventable.

Congenital abnormalities. These could be reduced by pre-pregnancy counselling — for example, to improve diabetic control or give vitamin supplements to prevent spina bifida — or by prenatal diagnosis followed by termination of an affected pregnancy. Prenatal diagnosis is now offered routinely in antenatal clinics in the form of an ultrasound scan, usually at around 19 weeks' gestation, to detect fetal anomalies. The range of anomalies that can be detected is steadily increasing with improvements in ultrasound expertise and equipment, but a district hospital may not match the standards set by tertiary referral centres, particularly with regard to cardiac abnormalities.

A woman who has an abnormal baby may blame the hospital for not offering her the appropriate test or referring her to a centre with more expertise. It has been suggested that blood testing for Downs syndrome should be offered to all pregnant women (Wald, 1990), but the decision to do so has financial implications for Health Service hospitals, and also has ethical implications. A hospital may decide not to offer testing to all women and this decision may be made by a committee of doctors or managers or both.

Antepartum causes of stillbirth (death *in utero* before labour).This sometimes has a specific cause such as maternal diabetes or infection. Maternity hospitals have a protocol of tests which are carried out on the baby and the mother when a stillbirth occurs, but in spite of this a cause may not be identified. Growth retardation may be recognized after delivery but as mentioned above the clinical diagnosis of growth retardation before delivery can be difficult.

Intrapartum stillbirth. This has become uncommon in present day practice in Britain but occasional cases still occur, and it has been suggested that a rate of one in every 1000 deliveries is an 'irreducible minimum'. Human error is often blamed for the loss of a baby which enters labour alive and apparently well. There may be failure to recognize abnormalities on a cardiotocograph trace which seem glaringly obvious in retrospect. Nevertheless, some obstetric disasters, such as amniotic fluid embolism, are difficult to predict and may happen so quickly that they cannot be treated.

Handicap. Of more importance as far as litigation is concerned are the possible causes of handicap, in particular mental handicap. Again, some are genetic, some are due to complications during the pregnancy, and some may be caused by intrapartum hypoxia or birth injury. It is now being recognized that a large proportion of mental handicap cases are due to genetic causes, that is, the problem lies in abnormal development of the brain not caused by pregnancy complications. Prenatal diagnosis has only a limited role to play here: for example, counselling before pregnancy may detect a risk due to consanguineous marriage, and tests during pregnancy can detect Down's syndrome, but may types of mental handicap cannot be detected by tests during pregnancy. Birth injury due to forceps can cause handicap if intracranial bleeding occurs, but direct injury from instruments is unlikely to be a cause of mental handicap in the absence of such bleeding.

In the past it has been too easy to blame intrapartum hypoxia for causing mental handicap or cerebral palsy. In many normal labours there may be an abnormality on the cardiotocograph trace, and if the abnormality was not acted upon and the child is handicapped the presumption has been that the court will connect the cardiotocograph abnormality with the child's subsequent condition and will award damages to the child (Symonds, 1988). However, as criteria such as those of Freeman and Nelson (1988) are being increasingly applied it is being recognized that most cases of cerebral palsy are due to causes other than asphyxia (Lamb and Lang, 1992).

Typical course of a complaint about care

Obstetric cases that lead to litigation often follow a depressingly similar pattern, and before discussing how the problem of litigation is to be tackled it may be of interest to describe the mythical but 'typical' case of Ms Smith.

Ms Smith attended early in pregnancy for antenatal care and saw her consultant at her first visit to the hospital clinic. She never saw him again. Subsequent hospital visits were frequent and she saw a different doctor on each occasion. No abnormalities were detected. She was admitted in spontaneous labour which at first appeared to be proceeding normally. She was cared for by a midwife and seen briefly by a junior doctor whom she had never seen before. Labour progressed slowly and the fetal heart was monitored electronically. Some minor abnormalities appeared on the trace and were commented on by the midwife in the midwifery notes. No routine ward round was done by the medical staff and no mention of the trace was made in the medical notes.

As the labour was progressing slowly the midwife called a doctor, and because of a change in shifts a different junior doctor saw Ms Smith. He discussed the case over the telephone with the registrar, and allowed labour to proceed. By now it was about 4 a.m. The cardiotocograph trace showed further abnormalities, but because the registrar had already been telephoned the midwife and the junior doctor allowed labour to continue rather than disturbing the registrar again.

Eventually the signs of fetal distress become even worse and the registrar — an overseas graduate — was called. He had never seen Ms Smith before. He expertly performed a forceps delivery but the baby was in poor condition. The baby spent some days in the special care nursery, during which time Ms Smith and her partner visited him regularly and heard the staff referring to 'birth injury'. Still uncomfortable after the delivery, Ms Smith became convinced that the forceps had injured the baby's brain. She was discharged from hospital without seeing the consultant responsible for her care.

The parents began legal proceedings against the hospital, and the hospital managers asked for a report from the consultant obstetrician. He reported that he saw her only once during the antenatal period, and did not comment on the management of the labour. By the time the registrar was asked for a report he had returned overseas. The junior doctors had also moved on to other posts. Several years after the birth the papers were sent to an expert assessor, who had to decide whether the management fell below the standards applicable several years previously (Drife, 1989a).

Problem areas

This fictitious case illustrates the problems identified by Ennis and Vincent (1990) in their review of 64 cases that had come to litigation over stillbirth, perinatal or neonatal death, or other problems. Three main topics of concern emerged in that study: inadequate fetal heart monitoring, mismanagement of forceps, and inadequate supervision by senior staff. On some occasions staff were unsympathetic and gave too little information (Vincent *et al.*, 1991).

Murphy *et al.* (1990) carried out a study in which the intrapartum cardiotocograph records of severely asphyxiated babies were compared with those of healthy infants. Investigators unaware of the clinical outcome agreed that abnormalities were present in the traces of 87 per cent of the asphyxiated infants and 29 per cent of the controls, and were severe in 61 per cent of the asphyxiated infants and 9 per cent of the controls. Fetal blood sampling was indicated in 58 per cent of cases in the asphyxia group but was actually carried out in only 16 per cent. The response of staff to the abnormalities was slow and the authors of this study concluded that 'the interpretation of cardiotocograph records during labour continues to pose major problems for practising clinicians'. In a study of the training of obstetric senior house officers in teaching hospitals and district general hospitals, Ennis (1991) found that most of these doctors received only one or two hours' teaching or lectures a week and some received even less. Half of them had had no formal training in interpreting or recognizing abnormal or equivocal cardiotocograms. When they were questioned at the end of their jobs about training in the use of forceps, 23 per cent said they had had no training, and 35 per cent of the remainder thought their training had been less than adequate. In a study of GP trainees' views on hospital obstetric training, Smith (1991) found that less than 40 per cent believed at the end of their six months that they were competent to perform a simple forceps delivery. Most of those questioned believed that more than six months' hospital training was necessary for a general practitioner who wished to provide care in labour.

The way ahead

'The real answer to the question "How to avoid medico-legal problems in obstetrics and gynaecology" is good practice and good communication' (Clements, 1991). Several improvements in practice are now being proposed.

Consultant involvement. The Royal College of Obstetricians and Gynaecologists has recommended that new consultant contracts should in future include sessions on the delivery suite. The Department of Health intends to reduce the long hours worked by junior doctors in 'hard-pressed' specialties and new consultant posts are being created, some of them in obstetrics and gynaecology, to reduce the workload on the juniors. It will be far from straightforward to ensure that these initiatives actually lead to changes in working practices in hospital, as the pressures on consultants to delegate labour ward duties will continue. The current NHS reforms may lead to closer scrutiny of doctors' work patterns, and this may enable consultants to avoid being drawn away from the delivery suite by other duties.

Training of juniors. The need for better training of senior house officers is becoming glaringly apparent, and as NHS managers become more aware of the importance of risk management, pressure to improve training will increase. With the NHS reforms, resources for training are being identified and should in future be better directed. There has been excessive complacency in British hospital practice that learning by osmosis is adequate for junior doctors, but the recent studies reviewed above have revealed how far training is falling short of what is needed. For example, it has been suggested that attitudes towards general practitioner training should change and that vocational trainees who wish to contribute to intrapartum care should be specially trained to do this (Pogmore, 1992).

Training of midwives. Schemes are being introduced to allow antenatal care to be shared appropriately between midwife, hospital clinic, and general practitioner. In the delivery suite, it seems likely that the use of electronic fetal monitoring will continue even if midwives gain more autonomy and are enabled to run sections of the delivery suite without doctors in attendance. Many midwives still do not feel comfortable with the interpretation of cardiotocograph traces and better education of midwives is required to teach them which types of pattern require further investigation.

Communication with patients. The importance of a good rapport with the woman and her partner is now recognized, and communication is being given a higher priority by doctors as well as midwives. 'The best protection for the doctor remains the one of talking to the patient and recording an outline of what is said.' (MacDonald, 1987). Good communication is essential once a problem has arisen, but good rapport with clients throughout pregnancy and labour will create a sound basis for full explanations if anything goes wrong (Chamberlain and Orr 1990).

Relationships between professions. Ideally, labour should be supervised by an experienced midwife who has immediate support from an experienced doctor (Drife, 1988a). The place of the inexperienced doctor on the labour ward will become more and more that of a trainee, learning from senior doctors and midwives (Pogmore, 1992). In general, consultants and midwives have good working relationships, particularly in the private sector. To provide this level of cover in the NHS will require increased resources from Health Authorities, who will need to be educated that increased investment in providing experienced staff will save money in claims as well as providing a better service for women and their babies.

References

Argent, V. (1988). Failed sterilisation and the law. *British Journal of Obstetrics and Gynaecology* 95, 113–15.

Barron, S. L. (1991). Audit in obstetrics. *British Journal of Obstetrics and Gynaecology* 98, 1065–67.

Beard, R. W., Filshie, G. M., Knight, C. A., and Roberts, G. M. (1971). The significance of the change in the continuous fetal heart rate in the first stage of labour. *Journal of Obstetrics and Gynaecology of the British Commonwealth* 78, 865–81.

Brown, A. D. G. (1985). Accidents in gynaecological surgery — medico-legal. In *Litigation in obstetrics and gynaecology: proceedings of the fourteenth study group of the Royal College of Obstetricians and Gynaecologists* (ed. G. V. P. Chamberlain, C. J. B. Orr, and F. Sharp). RCOG, London.

Chalmers, J. A. and Chalmers, I. (1989). The obstetric vacuum extractor is the instrument of first choice for operative vaginal delivery. *British Journal of Obstetrics and Gynaecology* 96, 505–506.

Chalmers, I., Enkin, M., and Keirse, M. J. N. C. (1989). *Effective care in pregnancy and childbirth.* Oxford University Press.

Chamberlain, G. (1991a). Organisation of antenatal care. *British Medical Journal* 302, 647–50.

Chamberlain, G. (1991b). Vital statistics of birth. *British Medical Journal* 302, 178–81.

Chamberlain, G. and Orr, C. (eds.). (1990). *How to avoid medico-legal problems in obstetrics and gynaecology.* RCOG, London.

Clements, R. V. (1991). Litigation in obstetrics and gynaecology. *British Journal of Obstetrics and Gynaecology* 98, 423–26.

de Regt, R. H., Minkoff, H. L., Feldman, J., and Schwarz, R. H. (1986). Relation of private or clinic care to the cesarean birth rate. *New England Journal of Medicine* 315, 619–24.

Department of Health (1989). *Report on confidential enquiries into maternal deaths in England and Wales 1982–1984.* HMSO, London.

Derom, R., Patel, N. B., and Thiery, M. (1988). Implications of increasing rates of caesarean section. In *Progress in obstetrics and gynaecology* Vol. 6 (ed. J. Studd). Churchill Livingstone, Edinburgh.

Drife, J. O. (1985). Operative delivery — clinical aspects. In *Litigation in obstetrics and gynaecology: proceedings of the fourteenth study group of the Royal College of Obstetricians and Gynaecologists* (ed. G. V. P. Chamberlain, C. J. B. Orr, and F. Sharp). RCOG, London.

Drife, J. O. (1988a). Sterilisation — the before and after. *Practitioner* 232, 39–43.

Drife, J. O. (1988b). Disciplining midwives. *British Medical Journal* 297, 806–807.

Drife, J. O. (1988c). My grandchild's birth. *British Medical Journal* 297, 1208.

Drife, J. O. (1989a). Doctors, lawyers and experts. *British Medical Journal* 299, 746.

Drife, J. O. (1989b). Professional conduct for midwives. *Midwife, Health Visitor and Community Nurse* 25, 410–12.

Ennis, M. (1991). Training and supervision of obstetric senior house officers. *British Medical Journal* 303, 1442–43.

Ennis, M. and Vincent, C. A. (1990). Obstetric accidents: a review of 64 cases. *British Medical Journal* 300, 1365–67.

Ennis, M., Clark, A., and Grudzinskas, J. E. (1991). Change in obstetric practice in response to fear of litigation in the British Isles. *Lancet* 338, 616–18.

Freeman, J. and Nelson, K. (1988). Intrapartum asphyxia and cerebral palsy. *Paediatrics* 82, 240–49.

Friedman, E. A. (1986). The obstetrician's dilemma: how much fetal monitoring and cesarean section is enough? *New England Journal of Medicine* 315, 641–43.

Grant, A. S. (1981). Some things never change. *Midwives Chronicle* 94, 348–50.

Hall, D. M. V. (1989). Birth asphyxia and cerebral palsy. *British Medical Journal* 299, 279.

Heringa, M. P. and Huisjes, H. J. (1988). Antenatal care; current practice in debate. *British Journal of Obstetrics and Gynaecology* 95, 836–40.

Hull, J. and Dodd, K. (1991). What is birth asphyxia? *British Journal of Obstetrics and Gynaecology* 98, 953–55.

Illingworth, R. (1985). A paediatrician asks — why is it called birth injury? *British Journal of Obstetrics and Gynaecology* 92, 122–30.

Johnson, N., Johnson, V., Fisher, J., Jobbings, B., Bannister, J., and Lilford, R. (1991). Fetal monitoring with pulse oximetry. *British Journal of Obstetrics and Gynaecology* 98, 36–41.

Kirkup, W. (1990). Perinatal audit: does confidential enquiry have a place? *British Journal of Obstetrics and Gynaecology* 97, 371–73.

Laguardia, K. D., Rotholtz, M. V., and Belfort, P. (1990). A 10-year review of maternal mortality in a municipal hospital in Rio de Janeiro: a cause for concern. *Obstetrics and Gynaecology* 75, 27–32.

Lamb, B. and Lang, R. (1992). Aetiology of cerebral palsy. *British Journal of Obstetrics and Gynaecology* 99, 176–77.

Leveno, K. J., Cunningham, F. G., Nelson, S., Roark, M., Williams, M. L., Guzick, D., Dowling, S., Rosenfeld, C. R., and Buckley, A. (1986). A prospective comparison of selective and universal electronic fetal monitoring in 34,995 pregnancies. *New England Journal of Medicine* 315, 615–19.

MacDonald, R. R. (1987). In defence of the obstetrician. *British Journal of Obstetrics and Gynaecology* 94, 833–35.

MacDonald, D., Grant, A., Sheridan-Pereira, M., Boylan, P., and Chalmers, I. (1985). The Dublin randomized controlled trial of intrapartum fetal hearth rate monitoring. *American Journal of Obstetrics and Gynaecology* 152, 524–39.

McKee, M., Priest, P., Ginzler, M., and Black, N. (1992). Can out-of-hours work by junior doctors in obstetrics be reduced? *British Journal of Obstetrics and Gynaecology* 99, 197–202.

Murphy, K. W., Johnson, P., Moorcraft, J., Pattinson, R., Russell, V., and Turnbull, A. (1990). Birth asphyxia and the intrapartum cardiotocograph. *British Journal of Obstetrics and Gynaecology* 97, 470–79.

O'Driscoll, K. and Foley, M. (1983). Correlation of decrease in perinatal mortality and increase in caesarean section rates. *Obstetrics and Gynaecology* 61, 1–5.

O'Driscoll, K. and Meagher, D. (1980). *Active management of labour.* Saunders, Eastbourne.

Paintin, D. B. (1990). Effective care in pregnancy and childbirth. *British Journal of Obstetrics and Gynaecology* 97, 967–69.

Paterson, C. M., Chapple, J. C., Beard, R. W., Joffe, M., Steer, P. J., and Wright, C. S. W. (1991). Evaluating the quality of maternity services — a discussion paper. *British Journal of Obstetrics and Gynaecology* 98, 1073–78.

Pogmore, J. R. (1992). Role of the senior house officer in the labour ward. *British Journal of Obstetrics and Gynaecology* 99, 180–81.

Quam, L., Dingwall, R., and Fenn, P. (1988). Medical malpractice claims in obstetrics and gynaecology: comparisons between the United States and Britain. *British Journal of Obstetrics and Gynaecology* 95, 454–61.

Saunders, P. (1992). Recruitment in obstetrics and gynaecology: RCOG sets initiatives. *British Journal of Obstetrics and Gynaecology* 99, 538–40.

Scherjon, S. (1986). A comparison between the organization of obstetrics in Denmark and The Netherlands. *British Journal of Obstetrics and Gynaecology* 93, 684–89.

Smith, L. F. P. (1991). GP trainees' views on hospital obstetric vocational training. *British Medical Journal* 303, 1447–50.

Smith, L. F. P and Jewell, D. (1991). Roles of midwives and general practitioners in hospital intrapartum care, England and Wales, 1988. *British Medical Journal* 303, 1443–44.

Symonds, E. M. (1985). Litigation in obstetrics and gynaecology. *British Journal of Obstetrics and Gynaecology* 92, 433–34.

Symonds, E. M. (1988). Without due care. *British Journal of Obstetrics and Gynaecology* 95, 434–36.

Treffers, P. E. and Laan, R. (1986). Regional perinatal mortality and regional hospitalization at delivery in The Netherlands. *British Journal of Obstetrics and Gynaecology* **93**, 690–93.

Vincent, C. A., Martin, T., and Ennis, M. (1991). Obstetric accidents: the patient's perspective. *British Journal of Obstetrics and Gynaecology* **98**, 390–95.

Wald, N. (1990). Serum testing for Down syndrome. In *Antenatal diagnosis: proceedings of and RCOG study group* (ed. J. O. Drife and D. Donnai). Spinger, London.

WHO (1991). *Maternal mortality: a global factbook*. World Health Organisation, Geneva.

4

Errors and accidents in surgery

JOHN H. SCURR

Surgery is an art that has evolved over the years, with great advances being made with the advent of new technology. This new technology has brought with it a number of problems, in addition to the enormous benefits that these advances confer. Greater patient expectation following surgical procedures and a greater awareness that problems can occur have undoubtedly led to an increase in medical litigation.

A surgical procedure follows a series of events, including a full history and appropriate examination and investigations, leading to an accurate diagnosis. Having achieved a diagnosis, planned treatment can then be carried out. The purpose of this treatment is to achieve a result that is considered satisfactory by both the patient and the surgeon. Surgery is not an exact science and deviations from the planned course may be perfectly acceptable and may encompass standard clinical practice. However, errors in judgement and technique can occur, producing an unsatisfactory result for both the patient and the doctor which may be avoidable. In some situations though, the outcome cannot be planned and the results cannot be guaranteed. Some surgical procedures are associated with known risks; these risks, and the chances of a successful outcome, form an important part of the consent.

Until 1989 the majority of medical accidents involving litigation in the UK were handled by the Medical Defence Associations. In an annual report these associations provided details of important cases, both won and lost on behalf of the doctor, including sums for which the case settled. With escalating costs and rapidly rising insurance premiums, an alternative method of dealing with claims was introduced. The introduction of Crown indemnity, with Health Authorities empowered to settle cases, means that the Defence Associations now no longer routinely become involved in all cases. With increasing legal costs, many small claims are now being settled with no admission of liability. The complexity of the legal system is such that even if a Health Authority wins a claim, it can be faced with costs running into tens of thousands of pounds with no right of recovery. Out-of-court settlements with no admission of liability attract no publicity and important lessons from such cases are lost.

Surgical audit

Auditing has become an important teaching aid and will undoubtedly influence surgical practice. Information, including the results of certain surgical procedures, will become available and patients, through their GPs, should have their choice of surgeon. Recommendations by the Royal College of Surgeons have led to the introduction of surgical audit, and this has provided much information about surgical outcomes.

In 1990 death within 30 days of a surgical procedure was the subject of a national confidential enquiry into perioperative deaths (NCEPOD). Seven thousand clinicians participated in this survey, the cases being anonymously reviewed by peers. Of 18 817 deaths reported within thirty days of surgery, one fifth of the cases, 3485, were randomly assessed. In 74 per cent of the cases, complete surgical data were obtained. The report analysed the deaths, the level at which decisions were taken, and the expertise of the operating surgeon, and identified shortcomings.

The original CEPOD report, published in 1987, showed preoperative decision-making by consultants in 63 per cent of cases, whereas in 1990 this had increased to 89 per cent. More operating is also now being done by, or supervised by, consultants, increasing from 47 per cent in 1987 to 67 per cent in 1990. The report draws attention to major clinical problems, including deep vein thrombosis and pulmonary embolism, resuscitation, and the management of elderly patients with fractured necks of femur and disseminated malignant disease.

Research into medical accidents

Why do medical accidents occur? Are there patterns and are they preventable? The Surgical Research Society is the premier research society for surgeons, and yet up to 1991 not a single presentation was made relating directly to medical accidents. Studies in the USA suggest that less than one in ten medical accidents are subject to investigation or complaint. Before any serious surgical research can be undertaken, a proper database recording surgical procedures and untoward events, complications, poor outcome, and patient satisfaction, is essential. However, until we define what constitutes a medical accident, it will be difficult to record and analyse all these events, and to make specific recommendations.

Adverse effects following drug administration are now reported and investigated. Should all unsatisfactory outcomes following surgical procedures be reported and subject to peer reviews? Some surgical procedures would then be noted to have a high incidence of complications, which might then prove unacceptable. Given that there is often more than one surgical procedure to correct a problem, specific recommendations could be made. Every incident involving an aircraft is subject to an enquiry by the Accident Investigation Branch of the Civil Aviation Authority. These findings are then published and directives issued if appropriate. Similar directives could be made regarding surgical practice.

Each surgical operation should be considered as a separate procedure requiring specific skills. Airline pilots are subjected to regular practical examinations and recency checks. A surgeon has no practical check, and certainly no recency check. In one recent case, a surgeon was performing a major surgical procedure on the oesophagus, having not performed this operation for over five years. Without knowing details of the patient outcome, one can only guess at whether this is satisfactory. In a recent negligence case, a patient was due to undergo corrective surgery, performed by a surgeon who had never carried out this procedure before — is that acceptable? Further important questions concern working conditions. Does tiredness affect surgical performance, does it affect clinical decisions, and should we restrict the number of hours that a surgeon works? Are the disadvantages of tiredness overcome by the benefits of continuity of care? No attempt to address these questions in a scientific manner has yet been undertaken.

Patterns are beginning to emerge with regard to medical negligence claims. Failure to communicate and adequately inform a patient about a surgical procedure is a common complaint. This trend has already been recognized and steps taken to provide better information prior to a surgical procedure. Patient preparation and preoperative investigation is sometimes inadequate. Problems here could be identified and corrected. Patient expectations may be quite unrealistic, both in terms of scarring and pain. Where scarring and pain are known to accompany the surgical procedure, care should be taken to explain to the patient so that they understand what outcome they are likely to expect. Patient dissatisfaction provides the basis for many claims, often compounded by the dismissive attitude of the surgeon, who feels that he has performed a great operation.

Perhaps the most interesting area of surgical research with regard to medical accidents relates to the assessment of surgical performance, and the regulation of surgeons who perform surgical procedures. Regulations which restrict clinical freedom may limit surgical options to the detriment of patients. The introduction of regulations could prove expensive and would not necessarily reduce the number of accidents. Operations by junior surgeons are frequently blamed in medical accident cases, but in reality, many of the accidents are caused by experienced surgeons. Delegation is important in any training programme, but the effect of delegation needs to be monitored carefully.

Preoperative preparation

Before undergoing surgery, all patients require preparation. This preparation will involve making an accurate diagnosis, assessing the operative risks to the patient, and advising about operative and postoperative complications.

Surgery can be conveniently divided into elective and emergency surgery. With elective surgery, a full history, a proper examination, and appropriate investigations should be carried out. The findings should be recorded, the results of investigations checked, and further investigations ordered to confirm or refute the diagnosis. Having achieved a diagnosis, the patient's fitness for surgery needs to be assessed. The importance of this assessment increases in patients with concurrent medical problems and advancing age.

To embark upon an elective surgical procedure without first establishing the patient's fitness to survive both the surgical procedure and the anaesthetic is not in accordance with good surgical practice. The purpose of an investigation is to assist in management and the choice of investigations is important. There can be no justification for over-investigating patients. The use of invasive investigations which do not assist diagnosis or management, but yet run the risk of complications, should be discouraged. Failure to check the result of an investigation may lead to a significant error in clinical management, and result in litigation.

In the emergency surgical case, great reliance may be made on clinical findings. An initial assessment, followed by resuscitation, may allow time for further investigations. In life-threatening states, requiring urgent surgical intervention, a less-experienced surgeon, who happens to be available, may have to work by clinical judgement rather than sophisticated investigations. An illustration of such a case involves a young man who was stabbed in the abdomen and who 'died' twice in the ambulance *en route* to hospital, but was successfully resuscitated and operated on by a junior doctor. His life was saved, but the doctor was unable to save his leg, whose blood supply had been compromised. A more experienced surgeon might have saved his leg, but to wait until he or she was available would undoubtedly have compromised the patient's life.

The operation

Surgery is an acquired skill. The more often the surgeon performs a particular surgical procedure, the better he or she becomes. All surgical procedures include a learning curve and to minimize the risks to patients, these operations should always be performed with a senior, supervising consultant present. However, surgical skills do not last forever, and with advancing age, some surgeons lose their acquired skills, or lack the ability to respond rapidly to a surgical situation.

Surgical practice can change overnight. We have, in the last three years, seen endoscopic surgery mushroom. The laparoscopy, widely used by gynaecologists, is now being used by many general surgeons for the purposes of cholecystectomy and other intra-abdominal and intrathoracic procedures. This revolution was brought about by technical advances in instrumentation, allowing these procedures to be performed. These procedures are associated with a number of risks. We have moved towards specialization and are now moving towards super-specialization. Thirty years ago the surgical disciplines divided nicely into general surgery, orthopaedic surgery, and specialist surgery, such as ENT, ophthalmology, and plastic surgery. We are now seeing the emergence of surgeons who specialize in vascular surgery, in urological surgery, in gastroenterological surgery, breast surgery, and many other separate disciplines. The justification for this super-specialization is not difficult to see, given advances in both diagnostic and technological procedures.

Attempts to minimize the risks of these new procedures by the organization of courses is to be commended. These patients have been associated with serious complications and fatalities. The use of newer technologies, including the laser ultrasound in the shattering of gall-stones, are again, associated with complications. Given the enormous benefits to patients of applying the technology, it is important that the risks are also explained and patients given a fully informed choice.

Postoperative monitoring

The purpose of postoperative monitoring is to note changes in the cardiovascular and respiratory system and respond appropriately. Routine operative monitoring now includes an electrocardiogram and pulse oximetry. For more complex surgical cases, measurements of central venous pressure and intra-arterial blood pressure monitoring may be carried out. For certain specialized surgical procedures, other specific monitoring may be required. In the immediate postoperative period, a recording of blood pressure, pulse, and respiration rate is essential for all patients. It is now common practice to use either the ECG or pulse oximetry into the postoperative period, to monitor the patient's condition. Patients who have undergone more major surgery will continue to have their central venous pressure and intra-arterial blood pressure monitored.

The frequency of postoperative monitoring depends on the patient's condition. A patient whose condition is improving requires less frequent monitoring than a patient who is stable or continuing to deteriorate. In patients whose general state is improving, the frequency of monitoring may be progressively decreased until the patient is fit to return to the ward. Patients should not leave recovery when it is necessary to monitor them more frequently than hourly, unless they are returning to the Intensive Care Unit or a high-dependency area.

In a patient who is clearly deteriorating, remedial action should be taken. Failure to take action is clearly negligent. Under these circumstances, the interval between monitoring may even be decreased, making the observations more frequent.

Postoperative complications

Postoperative complications can be divided into two: those associated with a specific operative procedure and those complications common to any surgical procedure. Many good surgical procedures have been ruined by inadequate postoperative care. Some complications are avoidable by careful technique, and others by the use of prophylactic agents to prevent them occurring. Where there is good evidence of a risk of complications, and known methods of prophylaxis which will reduce these risks, failure to consider or to employ them should be considered negligent.

PULMONARY COMPLICATIONS

Following general anaesthesia, respiratory depression occurs. This may result in partial collapse of the lungs, requiring expansion in the immediate postoperative period. A chest infection can accompany a general anaesthetic, but these changes take time to develop. A chest infection may not manifest itself for several days. It should be recognized and treated appropriately. Failure to recognize and treat a chest infection does not constitute good medical practice, and may result in a significant deterioration in the patient's condition.

CARDIAC COMPLICATIONS

Following any surgical procedure, alterations occur in the cardiovascular system. A prolonged fall in blood pressure may result in permanent ischaemic change in the heart, leading to a heart attack. Even in the best controlled anaesthetics, alterations in blood pressure can occur and the risk of developing a heart attack is well recognized. This should be identified at an early stage, and appropriate action taken. The risk of heart attack increases in certain high-risk patients, or in patients who have had previous cardiac problems.

DEEP VEIN THROMBOSIS/PULMONARY EMBOLISM

Deep vein thrombosis (DVT) occurs because of immobility and the effects of surgery. It leads to the blood setting in the legs, causing local damage to the veins, which may not manifest for ten or twenty years. On occasions, fragments break off, travelling to the lungs, leading to a pulmonary embolism, which can prove fatal. This is one of the most common and frequent postoperative complications, which is now preventable in many patients.

In patients over the age of forty undergoing major surgical procedures (an operation lasting more than thirty minutes), the risk of developing a deep vein thrombosis is 30 per cent. There are now well-recognized ways of reducing the incidence of deep vein thrombosis, by the administration of small doses of anticoagulants, subcutaneous Heparin, or the use of graduated compression elastic stockings, or intermittent pneumatic compression. When a patient is admitted to hospital for a surgical procedure, some assessment of deep vein thrombosis and pulmonary embolism risk should be undertaken. If the patient is considered to be at moderate or high risk of developing a deep vein thrombosis, then some form of DVT prophylaxis should be applied. Failure to consider the risk and apply appropriate DVT prophylaxis is now considered negligent.

WOUND INFECTION

A wound infection can occur following any surgical procedure, but is more likely to follow some surgical procedures than others. Operations on the bowel or biliary tract carry a significantly greater incidence of wound infection. Patients undergoing these operations should receive prophylactic antibiotics which reduce the incidence of wound infection.

The administration of antibiotics is not without its own risks. The common risks include allergic reactions, but potentially more serious risks include long-term problems with the bowel, leading to a form of colitis. Given that there are risks, some surgeons are still reluctant to use prophylactic antibiotics for all cases of bowel surgery, but there is agreement concerning operations on the large bowel, the appendix, and potentially contaminated operations where the use of prophylactic antibiotics have been shown to be beneficial. Failure to use antibiotics under these circumstances would be negligent.

INCISIONAL HERNIA

An incisional hernia occurs when the wound partly disrupts; the deeper layers of the wound have become weakened during the postoperative healing. This may be associated with a recognized or unrecognized wound infection, undue abdominal distension during the healing phase, or general debility, reducing the patient's ability to lay down good-quality fibrous tissue.

The development of an incisional hernia could not be considered negligent unless it could be shown that inadequate steps were taken to prevent infection, the technique used to close the wound was inadequate and should not have been used, or that there were complicating factors in the postoperative period which made the patient prone to develop an incisional hernia. Most claims arising from the development of an incisional hernia have succeeded because of poor management of the hernia, rather than because of its occurrence in the first place.

BLEEDING AND HAEMATOMA FORMATION

Bleeding accompanies any surgical procedure. However, some patients have an underlying clotting abnormality. To operate on a patient with a past history of bleeding problems, without investigating these and without having adequate supplies of blood ready, could prove disastrous. Occasionally, serious and unexpected bleeding occurs during any surgical procedure. However, even when a major artery is injured during the course of surgery the situation is remediable. Failure to control such a situation usually reflects an inexperienced surgeon, operating without adequate supervision. Some bleeding is unpredictable and not caused by any negligent act; failure to recognize bleeding and take appropriate steps can, however, be criticized.

Future research

Given that we know that a large number of medical accidents occur, there must be scope for fruitful research. The key to any research programme is accurate data collection and addressing the correct questions at the outset. Information concerning claims for medical accidents should be recorded. The Royal College of Surgeons is now obtaining data on patients who die during or shortly after operations. For this information to be valuable, it is important that all the data are recorded, that every death is reported, and a full analysis of the circumstances carried out. A regular report will then provide information to surgeons which will help them to prevent avoidable deaths.

The majority of medical negligence cases do not involve deaths, but patients who have survived with injuries. Although an increasing number of surgeons keep their own audit, and indeed are encouraged by the Royal College of Surgeons to do so, some central collation of this data would be an advantage. Many complications go unreported and indeed many do not form the basis for a medical negligence claim. Information concerning claims for medical accidents should be recorded.

Attempts to identify frequent claims should be undertaken, and subject to a proper investigation. This information should be made available to surgeons who should then incorporate it into their clinical practice. Given that some accidents will always occur, early recognition and compensation should be undertaken and better methods of assessing patient suffering should be explored. In an attempt to prevent further accidents, the circumstances surrounding the accident should be widely reported. If we know, for example, that the risk of deep vein thrombosis and pulmonary embolism can be reduced by prophylactic agents, all patients should at least be considered for these agents. Clinical circumstances may make it appropriate to apply a specific prophylactic method, but if that were the case, specific reasons should be listed. Failure then to consider the risk would prove negligent and attract automatic compensation.

Summary

Errors and accidents will continue to occur in surgery. The full extent is unknown, but should be subject to careful record-keeping and analysis. To prevent errors being compounded, accurate reporting and dissemination of this information should be freely available. Errors and accidents can occur during the investigation and treatment of patients without adversely affecting the outcome. A combination of errors and a failure to recognize a problem and to respond appropriately is more likely to lead to a poor outcome and hence litigation. An understanding of what makes patients sue doctors should assist in reducing medical litigation and improving both the practice of surgery and the quality of the outcome.

References

Campling, E. A., Devlin, H. B., Hoile, R. W., and Lunn, J. N. (1992). *Report of the National Confidential Enquiry into perioperative deaths, 1990.* National Confidential Enquiry into Perioperative Deaths, London.
Jackson, J. P. (eds.) (1991). *A practical guide to medicine and the law.* Springer, Berlin.
Powers, M. and Harris, N. (1990). *Medical negligence.* Butterworths, London.

5

Errors and accidents in anaesthetics

MICHAEL WILSON

Introduction

Interest in accidents and errors in anaesthetics has greatly increased in the last
two decades. There are now many excellent publications, whereas twenty
years ago only sporadic letters appeared in the relevant journals. Interest has
shifted from ergonomic detail to the psychological concepts of linkage (Gaba
et al., 1987), latent error (Eagle *et al.*, 1992), and human performance
(Wilson, 1988; Gaba, 1989; Weinger and Eglund, 1990). Much of the
impetus for this has come from the work of the Anesthesia Patient Safety
Foundation in the USA.

The administration of an anaesthetic does not always run according to
plan and unexpected incidents are common. Although such incidents seldom
have serious consequences they may sometimes cause harm if they are
undetected or if the anaesthetist fails to deal with the resulting emergency;
they would then usually be called accidents. Incidents may be conveniently
classified as being within or without the anaesthetist's control. Incidents that
occur outside the anaesthetist's control happen quite unexpectedly. The
anaesthetist is not at fault, but he or she is responsible for preventing any
potential harm to the patient. Two examples of such incidents are the failure
of a ventilator and an anaphylactoid drug reaction. Incidents due to human
failure are potentially within the anaesthetist's control. They involve definite
errors, for example, the incorrect setting of gas flows or a failure to notice
cyanosis. Errors of this kind are often referred to as slips or lapses of performance but if errors arise because of a failure to apply accepted rules or
knowledge correctly they are usually termed mistakes (Reason, 1990,
Chapter 1). Some incidents are considered avoidable because they result from
a 'departure from accepted practice' that is, they are due to substandard care
or even negligence. Some of these incidents may be due to error, possibly
ignorance or faulty judgement.

If an incident goes undetected it may lead to an unstable situation which
ultimately may cause harm. However, often an incident will not lead to
serious injury. Some incidents, although undesirable, are unlikely to result in
much damage; other incidents may have a higher likelihood of causing
serious injury but this is not an inevitable outcome for several reasons:
(a) first and foremost, the body is remarkably resistant to physical insult — it

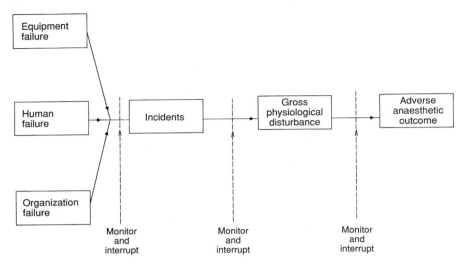

Figure 5.1   Classification of anaesthetic incidents and their consequences (from Wilson, 1985).

can tolerate extreme conditions for brief periods (e.g. profound hypoxia or hypotension) and it can invoke compensatory reflexes to restore homeostasis; (b) the incident may be too short-lived to cause any major effect; (c) the anaesthetist should recognize that something is amiss and take urgent corrective steps to interrupt the process and recover the situation (Gaba, 1989).

Accidents, incidents, and errors usually have causes which one hopes to identify so that preventative action can be taken. An investigation may reveal certain immediate causes but the causal chain will sooner or later (usually sooner) come to an unsatisfactory halt. Often a variety of factors contributes to a harmful event; Galletly and Mushet (1991) suggest that an average of 8.1 factors may be involved. Some factors, which alone are insufficient to cause harm, may nevertheless contribute to the final event (e.g. if a disconnection, which is harmless if noticed, is missed because of fatigue). On other occasions an incident may 'propagate', with one error leading to another, causing a chain of incidents that may get completely out of control. Perversely, a change in anaesthetic technique or equipment design introduced to eradicate a known error may trigger a new incident with possibly even worse consequences (Cooper *et al.*, 1978; Keats, 1990). Also the very act of recovering from even a harmless incident may create an even more serious error if the anaesthetist makes the wrong response (DeAnda and Gaba, 1990).

The classification of anaesthetic incidents lacks an agreed scheme and is far from satisfactory. One suggestion (Wilson, 1985) is that all incidents are due to human failure, equipment failure, organizational failure, or a combination of these factors (see Fig. 5.1).

An incident that may harm the patient is referred to as an adverse anaesthetic event. This, if undetected, may result in some gross disturbance of physiology which if it is also not detected may cause death or injury (adverse anaesthetic outcome or anaesthetic-related complications). Damage may be averted at several points along the chain of events if some form of monitoring detects the aberration, allowing the anaesthetist to intervene and recover the situation. Similar models that also emphasize the interruption of the accident cascade are presented by Gaba *et al.* (1987) and Galletly and Mushet (1991).

Surveys

The frequency of errors and accidents in anaesthesia which cause death is difficult to determine because the sequence of events that cause a death is often unclear, and the most important factor is usually the gravity of the patient's condition. In the UK a National Confidential Enquiry into Perioperative Deaths (NCEPOD) reviews clinical practice to identify remediable factors in the practice of anaesthesia and surgery. In the first report (Buck *et al.*, 1987) each reported perioperative death was assessed by an anaesthetist and a surgeon in an attempt to decide if the death was due to disease, surgery, or anaesthesia. Anaesthesia was held to be partly responsible for 1 in 1300 deaths and was completely responsible for 1 in 185 000 deaths. Since then, regular national studies have continued but both the earlier use of assessors and the attempt to determine avoidability or culpability have been abandoned (Campling *et al.*, 1992). Accidents and errors are only dealt with in tables of 'untoward incidents'. The chief value of these reports is in the strongly worded conclusions that call for specific improvements in the delivery of anaesthesia. Mortality surveys from many other countries have been published: in all studies, common causes of death recur (see Table 5.1).

Probably the most reliable published surveys are the triennial reports on Confidential Enquiries into Maternal Deaths in the UK. All maternal deaths are reported and closely examined by obstetric and anaesthetic assessors. Unfortunately it is not yet possible to provide an accurate estimate of the incidence of maternal deaths attributable to anaesthesia because the total number of obstetric anaesthetics given is not known. In the most recently published report covering the years 1985–87, eight mothers died as a direct result of the anaesthetic (a reduction compared to previous years). Six of these deaths were due to failure to detect that the tracheal tube was misplaced or kinked. In addition, anaesthetists were held to have contributed to 16 other deaths (Department of Health, 1991).

Table 5.1   The common causes of anaesthetic deaths
(reproduced with permission from Wilson, 1985)

Gross disturbances of the respiratory system
   Inhalation of gastric contents
   Intubation problems
   Hypoventilation and airway obstruction
Gross disturbances of the cardiovascular system
   Unrecognized hypotension — hypovolaemia
                       — drug overdose
Mechanical failure
   Failure of oxygen supply
   Leaks
   Disconnection
   Obstruction
Human failure
   Inadequate training
   Error
Organizational failure
   Supervision and assistance
   Recovery facilities
Monitoring
   Inadequate monitoring

Another approach is to examine the records of the medical defence societies since, for their own protection, anaesthetists involved in any unfortunate events are likely to report the details (Aitkenhead, 1991). These sources cannot provide the absolute incidence because the number of anaesthetics given by the members of a particular society is unknown. Individual cases of special interest are sometimes reported in detail (e.g. in the series 'Anaesthesia and the law', Brahams, 1989). In the most recent UK review, Gannon (1991) analysed 25 deaths reported to the Medical Protection Society's London office during the period 1982–86. Three of the more common problems related to intubation, use of drugs, and use of equipment. The two most common associated factors were inadequate supervision of juniors and inadequate preoperative assessment.

In the USA the Committee on Professional Liability of the American Society of Anaesthesiologists has been studying closed malpractice claims since 1985. Their findings have been published in a series of reports with twenty insurance companies throughout the USA now participating. A standardized form is used to record all information and each claim is reviewed by a practising anaesthetist according to a detailed set of instructions. Reviewers write a brief report and, referring to current standards, assess the adequacy of care and its contribution to the outcome. The Closed Claims Committee, comprising three practising anaesthetists, approves the on-site reviewer's assessment.

The most common harmful anaesthesia-related outcomes in the American series were death (37 per cent), nerve damage (15 per cent), and permanent brain damage (12 per cent) (Cheney *et al.*, 1989). The standard of care was considered inappropriate or substandard in 50–80 per cent of claims, depending upon the type of incident. There was evidence of short cuts, inadequate monitoring, serious errors of judgement, and poor choice or conduct of the anaesthetic. Respiratory events constituted the single largest source of claims (Caplan *et al.*, 1990). Thirty - eight per cent of these claims were accounted for by inadequate ventilation with no specific identifiable cause, 18 per cent caused by oesophageal intubation, and 17 per cent caused by difficult tracheal intubation. Better monitoring might have prevented three-quarters of the instances of inadequate ventilation and oesophageal intubation. An important conclusion from these reports is the failure of routine clinical methods to identify the correct position of the tracheal tube.

There are some major difficulties in using these surveys to provide information about accidents and errors, especially their incidence. These are associated with definitions, assignment of cause, and classification. Death is an outcome easy to define but morbidity, which may include anything from a bruise to brain damage, presents greater problems. Even the definitions of an incident or a reasonable standard of care reflect widely differing opinions. The critical reader should also be aware, however, that one can have little confidence in many of the estimates of incidence because they are sensitive to very small numerators. For example, the often-quoted risk of anaesthesia is 1 death in 185 000 anaesthetics, but this is based on only three deaths in which anaesthetic management was judged to be the sole cause of death (Buck *et al.*, 1987).

A second major difficulty is the assignment of cause (Keats, 1990). A death or complication may be due to the anaesthesia, the surgery, the nursing care, the patient's disease, a chance event, or a combination of some or all of these factors. Accidents, errors, lack of knowledge, lack of judgement, or lack of skill may all contribute. The task of making these assignments is usually left to one or two assessors who are provided with a report or the original notes, which are often indecipherable or incomplete. The cause may be obvious but sometimes it is impossible to separate the various contributions or to claim that anaesthesia was totally or even partially responsible for a death. Inevitably the judgements will be influenced by hindsight (Keats, 1990). Consequently one must expect considerable variability between assessors' judgements. Recent work also indicates that these judgements are biased by knowledge of the outcome of the incident (Caplan *et al.*, 1991). An interesting new technique is to require judgement to be passed sequentially at each stage as details of an incident are unfolded (Cook *et al.*, 1991).

The reliability of assessors' judgements has been given scant attention and the extent of the assessor's agreement (or disagreement) is not documented. It is of interest that 'In CEPOD, assessments of the clinical care of patients were made by a large panel of assessors chosen broadly from the different specialities. This method of peer review was cumbersome and difficult to validate, and has subsequently been criticised. Critics have alleged that the peer reviewers chosen used unrealistic standards to measure the process of clinical care and often expected unobtainable standards for the average district hospital consultant to achieve.' (Campling *et al.*, 1992). Not surprisingly, NCEPOD has dropped this form of assessment. How the assessors now function and how well they agree is not clear.

The authors of the Closed Claims Study of the American Society of Anesthesiologists are to be admired for the way they have tackled this technically difficult problem. In a first study they presented 42 anaesthetists with details of 48 mishaps of varying severity of injury and quality of care (Caplan *et al.*, 1988b). The anaesthetists were asked to make three judgements — adequacy of care, the type of human error (mistaken judgement, technical error, or lapse in vigilance), and whether better monitoring would have prevented the complication. They concluded that 'anesthesiologists from widely different backgrounds can produce a relatively cohesive set of judgements when asked to review anesthetic mishaps'. However, this conclusion is scarcely supported by their finding that '...one-fourth of the participants disagreed with their peers on each issue', and this despite providing information on a standardized ten-page data collection form and using an instruction manual. When explicit instructions were not provided, anaesthetists failed to reach an accepted level of agreement (Posner *et al.*, 1991).

It is important to consider what is required of these attempts to measure agreement. The ideal would be a perfect judge assessing each report and giving a judgement representative of the average anaesthetist and devoid of bias and variability. In the real world one must rely upon judges who approximate to this ideal. This places an onus upon those who use assessors to show that they are in good agreement with each other (and can be regarded as representative). How much deviation from the ideal can be accepted is a matter of opinion (as is the more familiar statistical significance level for rejecting an hypothesis).

Finally, there is no agreed system of classification in the various studies of anaesthetic misadventure. Some classifications of adverse anaesthetic events are a mixture of mechanisms (e.g. failure to intubate), physiological responses (e.g. arrhythmias), and final outcomes (death or injury). There is often overlap so that an event could be listed under one or all of several categories.

Different studies also use different numbers of categories in their classification. Relative frequencies of different events need to be examined closely to see how they have been derived because they are dependent on the number of categories used in the classification.

CRITICAL INCIDENT STUDIES

Large studies are disappointing for those interested in the causes of accidents and errors. They do not provide detailed information or the causal chains that lead to each injury. Cooper *et al.* (1978) in the USA suggested that it might be more profitable to study incidents rather than harmful outcomes. Incidents are commonplace and therefore a better starting point for a study than the rare event of death. Incidents are also worth studying even when no harm ensues because sooner or later they may trigger a sequence of events leading to injury. Such incidents are called 'critical incidents' and through a fortuitous double meaning the term is particularly appropriate since it suggests an impending crisis. For Cooper's purpose a critical incident was defined as '...a [preventable] occurrence that could have led (if not discovered or corrected in time) or did lead to an undesirable outcome, ranging from increased length of hospital stay to death or permanent disability'.

The first study (Cooper *et al.*, 1978) was a retrospective one in which 47 anaesthetists were interviewed and asked to recall their critical incidents. This study drew on years of experience and many thousands of anaesthetics, but has obvious biases. In particular, instead of reporting all incidents, anaesthetists may selectively report mainly those they considered important. The technique was then developed to include prospective reports from trained observers who were interviewed over a telephone 'hot-line' immediately after an incident (Cooper *et al.*, 1982), and a series of important papers followed, all stressing the importance of human error in critical incidents. Originally, Cooper *et al.* (1978) arbitrarily classified disconnections as human errors, perhaps because anaesthetists should couple connectors carefully and maintain a vigilant watch over them. This decision inflated the relative proportion of human error to 82 per cent of all critical incidents. An alternative view is that disconnections are the result of a design fault and should be classified as equipment failure. In their 1984 paper Cooper *et al.* sat on the fence and created the separate category of disconnections (14 per cent) so human error was now responsible for about only 70 per cent of incidents and equipment failure for about 15 per cent of them.

By 1984 a 'library' of 1089 incidents had been collected with 70 of them having substantive negative outcomes (SNOs — death, cardiac arrest, or extended stay). Excluding disconnections, common sources of equipment failure were with the breathing system and the monitoring devices (mainly involving ECG leads or oscilloscopes) (Newbower *et al.*, 1980); common

Table 5.2 Strategies for potential prevention or detection (from Cooper *et al.*, 1984)

---

 1. Additional training
 2. Improved supervision/second opinion
 3. Specific protocol development
 4. Equipment or apparatus inspection
 5. More complete preoperative assessment
 6. Equipment/human factors improvement
 7. Additional monitoring instrumentation
 8. Other specific organizational improvements
 9. Improved communications
10. Improved personnel selection procedures

---

human errors were associated with drug administration, use of the anaesthetic machine, and management of the airway. Associated factors that might have contributed to the occurrence of critical incidents included failure to check equipment, a first experience with the situation, inattention or carelessness, haste, and unfamiliarity. Cooper *et al.* (1984), as a result of their analysis, recommended ten potential strategies for the prevention or detection of 70 incidents that had SNOs (Table 5.2).

The technique has also been used to study how anaesthetists detect a sudden disruption (breathing system disconnection) or a gradual deviation (hypovolaemia) (Newbower *et al.*, 1980). Half of these events remained undetected until unacceptable signs of patient distress were noticed. This little-publicized work is important because it provided evidence to challenge the traditional beliefs that the ECG and oesophageal stethoscope are reliable monitoring devices, and also the belief that one could rely upon the anaesthetist's vigilance.

The work from Cooper's group has been described in some detail because not only were they responsible for introducing a new investigative method to anaesthesia, but they also applied it in several important areas and provided a thorough analysis of preventable anaesthetic mishaps. Since the original publication several prospective studies have been published from, for example, the UK (Craig and Wilson, 1981), elsewhere in the USA (Kumar *et al.*, 1988), Canada (McKay and Noble, 1988), Australia (Currie, 1989), New Zealand (Galletly and Mushet, 1991), and the Netherlands (Chopra *et al.*, 1992).

However, after the initial promise, the results of critical incident studies have been a little disappointing. The chief difficulty is that many anaesthetists cannot be bothered to make reports, especially if they are unsure whether trivial incidents are worth reporting. Compliance rates of only 10–50 per cent (McKay and Noble, 1988; Currie, 1989) have been quoted but it is likely to be much less than this for the more trivial incidents. After some initial enthusiasm, the frequency of reporting diminishes although it may be sustained if the person supervising the collection makes vigorous attempts to maintain each individual's involvement. Feedback is essential and it is important to publicize incidents locally, discuss them at mortality and morbidity meetings, and demonstrate that they are instrumental in making changes to improve safety.

In theory, the frequency of critical incidents could be calculated if the number of anaesthetics given during the period of collection were known but the estimate would be very unreliable because of the high, but unknown, amount of under-reporting. It is also seldom possible to make meaningful comparisons between different studies because of the lack of an agreed method of classification. McKay and Noble (1988) point out that the critical incident method is primarily a problem-solving technique which cannot be relied upon for quantitative purposes. Perhaps the greatest disappointment is that because of uncertain co-operation, reporting rates are not reliable and stable enough to measure the effect on safety of the introduction of new equipment (Cooper *et al.*, 1987; but see Kumar *et al.*, 1988).

Task analysis

A completely different approach to safety is to examine the anaesthetist's task so as to identify potential sources of error and accident (Wilson, 1988; Gaba, 1989; Weinger and Englund, 1990). The following is a simplified schematic approach to understanding the human factors underlying errors in anaesthetics (Wilson, 1988).

The anaesthetist acquires information about the state of the patient and the progress of the anaesthetic, both from monitors and from observation of the patient. He or she compares this information with a mental model of how the anaesthetic should be proceeding, makes a decision to adjust the anaesthetic if necessary, and notes the effect (see Fig. 5.2). The anaesthetist's perception of the situation is subject to many influences, including expectations of what has usually been observed in a particular situation. Furthermore, it is known that selective attention leads to emphasis on particular sources of information; other signals will be kept waiting, stored possibly for later processing (see Fig. 5.3). The selective processing system is limited in capacity and may easily become overloaded (Gaba and Lee, 1991). For this reason some information, including perhaps important features of a situation, will be neglected. It is for this reason that distractions are potentially dangerous.

The anaesthetist continuously compares the information with a model of how the anaesthetic should be proceeding and consults a store of knowledge and experience. Various hypotheses about the meaning of the new information and the best course of action must then be evaluated. It is here that many factors, including personality traits, motivation, and level of arousal,

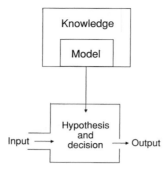

Figure 5.2 The anaesthetists's decision-making process.

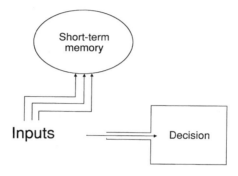

Figure 5.3 Selective attention

influence the anaesthetist's judgement and choice of action. Following a decision to change the anaesthetic, some action must be correctly executed. An error will result if a control is incorrectly adjusted, perhaps because of poor equipment design. The successful performance of such actions requires confirmatory feedback from the machine or the patient.

Analysing the anaesthetist's task in this manner has led to many major improvements over the last decade. The classification of errors into those that are design-induced and those that are operator-induced has been particularly useful as will be apparent later. In general, safety can be improved by reducing the number of errors, adopting safer techniques, and employing detection and recovery techniques.

ERROR REDUCTION

The obvious way to improve anaesthetic safety is to discover the causes of accidents and errors and take steps to prevent their recurrence. This is not as profitable as it might seem. Causal investigation may not penetrate beyond the obvious and will often not lead to the identification of critical causal factors. Furthermore, incidents may reflect the characteristics of a specific working environment, making it difficult to generalize to others. Nevertheless, Armstrong and Davies (1991) have developed a detailed protocol which may be useful for investigating anaesthetic incidents. This emphasizes chronological order rather than concentrating on patient and equipment factors.

There is another major problem — latent errors (Reason, 1990; Eagle *et al.*, 1992). Preventative strategies that depend on known errors cannot influence more deep-rooted organizational problems (see Chapter 1). They also cannot allow for the extraordinary actions of some anaesthetists or for the fact that some events arise because of an unusual combination of events, especially as anaesthesia becomes more complex (Gaba *et al.*, 1987).

Nonetheless, it is possible to suggest a number of safety measures on the basis of past incidents and an understanding of the anaesthetist's task and working environment. These will be considered under the headings previously used (Fig. 5.1): equipment, human behaviour, and the organization. In addition, it is important to consider the interfaces between humans and equipment, and between humans and organizations.

Equipment. Since faulty equipment rarely causes serious accidents (DeAnda and Gaba, 1990; Barton, 1991; Campling *et al.*, 1992) or critical incidents (Cooper *et al.*, 1978; Craig and Wilson, 1981; Cooper *et al.*, 1984) there is the risk of complacency despite reports of new faults appearing in most issues of anaesthetic journals. Breathing system disconnections are frequent and if undetected can cause death (Cooper *et al.*, 1978; Craig and Wilson, 1981; Cooper *et al.*, 1984; Thompson, 1987; Currie *et al.*, 1989). Because some component of the breathing system may disconnect several times a day, most anaesthetists accept this as a routine hazard. Why antidisconnection devices are not in general use is a mystery. The most likely reasons are inertia, low perceived risk, greater cost of the alternatives, and anxiety that a new type of connector will introduce its own hazards.

An important lead in improving equipment safety has been taken by the various national and international standards organizations (Thompson, 1987). In the UK the Medical Devices Directorate (MDD) (Barton, 1991) plays an important role in developing these standards, evaluating equipment, and approving manufacturers. The MDD is responsible for the investigation of accidents with medical devices and it issues Hazard Notices and Safety Action Bulletins to warn other users.

The Human/equipment interface. Over recent years there have been great improvements in the design of equipment to make it easier to use and less likely to be misread or set incorrectly (Thompson, 1987). Several factors are responsible, notably the application of ergonomic principles, and a responsiveness by manufacturers to criticisms from users and the MDD.

A very common cause of accidents is failure to understand the use of equipment (Cooper *et al.*, 1978, 1984; Craig and Wilson, 1981; Kumar *et al.*, 1988). This is no surprise as a survey revealed that 48 per cent of anaesthetists use new equipment without reading the manual (Weir and Wilson, 1991). Some manuals are ignored because they are excessively long and poorly written. Sometimes, because of urgency or other force of circumstance, there is no opportunity for training, and organizations commonly fail to ensure that sufficient time is devoted to training (and re-training!). Another very common reason is that reading a manual requires effort so an anaesthetist may resort to experimentation with the controls. Trial and error learning is a common preferred activity and it seems essential for learning many modern tasks (e.g. using a computer or a hired car). It may be wiser to accept this human weakness and make allowances for it in the design rather than expect to eradicate it completely by condemnation. Whenever possible, equipment should be simple to use, have sensible default settings, and depend upon a minimum of written instructions on the casing.

Another common cause of incidents is failure to check equipment before use (Cooper *et al.*, 1978, 1984; Craig and Wilson, 1981; Kumar *et al.*, 1988; Chopra *et al.*, 1992): 30–41 per cent of anaesthetists perform no checks and, of those that do, few follow the Association of Anaesthetists' guidelines (Mayor and Eaton, 1992); 60 per cent of anaesthetists do not follow the manufacturer's check procedure (Weir and Wilson, 1991). In the USA the Food and Drug Administration has recommended a lengthy check list yet, surprisingly, anaesthetists who were told to follow it found only 30 per cent of the prearranged faults in a modern machine (March and Crowley, 1991). Excuses included the complexity and length of some check lists as well as disagreement with their content. None of these is an acceptable reason for failing to follow some sort of checking procedure. It seems that the risk of serious injury is perceived as being so small that the effort is not seen to be justified. This view is strengthened by the knowledge that trivial equipment-related incidents are a daily occurrence and they are almost always detected and rectified — a very dangerous attitude.

Human failure. This is responsible for perhaps 60–80 per cent of anaesthetic incidents (Cooper *et al.*, 1978, 1984; Craig and Wilson, 1981; Kumar *et al.*, 1988; DeAnda and Gaba, 1990; Aitkenhead, 1991; Chopra *et al.*, 1992).

Modern technology, using the microchip and ergonomic design principles, has produced a new generation of anaesthetic monitors (Schreiber and Schreiber, 1987; Thompson, 1987). Their displays integrate the information, and changes from normal or preset values are easily noticed; a discrete sound can draw attention to a problem and its urgency (McIntyre, 1991), the nature of which is displayed visually on the monitor. The overall aim is to provide information that is easily assimilated, thus reducing the risk of information overload.

However, the modern anaesthetist, provided with all this information, is encouraged, if not obliged, to attend to more information. There is now more to worry about and more concentration is required. So paradoxically, the new equipment can be more tiring to use. Far more information is now offered to the anaesthetist than before and this allows incidents to be picked up at an earlier stage of their development than formerly. Unfortunately, false alarms are still very common with current monitoring equipment (McIntyre, 1991). This usually means repeating the observation and/or acquiring other confirmatory evidence (Gaba *et al.*, 1987, Gaba, 1989) by direct observation of the patient or from other monitors. When there is conflicting information older anaesthetists may rely on their own senses (e.g. the patient's colour and a finger on the pulse) but humans are known to be unreliable observers so this may not be a wise solution. The reliability of man *versus* machine in this context has not been examined nor have the relative merits of different display methods.

Gaba and Lee (1991), using a secondary substitution task as an indirect measure of work load, have shown that at times the capacity to attend to additional tasks is reduced. Distractions (including social chatter) can interfere with attention and decision making. Even legitimate tasks such as manually taking the patient's blood pressure can divert attention. Nowadays this problem can be overcome by employing automatic data collection. Automatic logging of the data eliminates the distraction of having to record data especially when the anaesthetist should be concentrating upon an unstable phase of the anaesthetic. However, it is not clear if there is further advantage in an automatic printout of the data (other than for medico-legal purposes), since it may be argued that making a manual record (when there is a lull) usefully maintains attention and encourages a review of the anaesthetic.

Traditionally, constant vigilance is expected (Newbower *et al.*, 1980; Gaba, 1989; Weinger and Englund, 1990). Yet any honest anaesthetist will admit that successful performance cannot be maintained all the time. This human fallibility has been confirmed by critical incident studies (Newbower *et al.*, 1980). Thus although vigilance is needed this must be done with the support of well-designed equipment monitors. Critical incident studies also suggest that there is an advantage in changing anaesthetists during a long case (Cooper *et al.*, 1982).

Human failure in decision making. The anaesthetist has to decide not only *if* something is wrong but also *why*. The monitors indicate which vital signs are abnormal (e.g. an unusually low pulse rate) leaving the anaesthetist, who is usually now under stress, to piece together the various items of information and choose a causal hypothesis (e.g. hypoxia or peritoneal traction) which will lead to corrective action. Usually the anaesthetist will follow some, perhaps unconscious, mental rule and choose the most common explanation that matches the situation. This is 'frequency gambling' (Reason, 1990) and although this will be most frequently correct, it may not always be the best strategy (DeAnda and Gaba, 1991).

Where a problem is completely new, abstract reasoning will be required for its solution. This is knowledge-based behaviour which is slower and requires more effort. In these circumstances anaesthetists need to be taught to question their decisions because clinging to false hypotheses or an inappropriate rule is a well-known cause of accidents. This dangerous form of 'keyhole' thinking (Reason, 1990; DeAnda and Gaba, 1990; Schwid and O'Donnell, 1992) is especially common in certain circumstances (Hawkins, 1987). Consequently, disastrous decisions may be adhered to, irrationally and tenaciously, despite conflicting evidence (e.g. the anaesthetist insists that a tracheal tube is correctly placed when, in fact, it lies in the oesophagus). Perhaps the remedy is to teach awareness of this danger and the need to listen to other opinions. In whatever way a decision is reached it must be verified ultimately by the patient's response.

Human/Organization interface. Fatigue, lack of sleep, hunger, frustration, excessive work load, and poor morale are all likely to decrease performance (Gaba, 1989; Weinger and Englund, 1990; Anon., 1992). Although experience and common sense supports this view it is hard to find satisfactory experimental justification and the number of incidents attributed to these factors in surveys is few (Craig and Wilson, 1981; Cooper *et al.*, 1984; Buck *et al.*, 1987). It is of grave concern that many anaesthetists work beyond their limits and recall making errors when fatigued (Gravenstein *et al.*, 1990). These issues are ultimately the responsibility of the hospital management, as is the duty to ensure that Hazard Notices and Safety Action Bulletins issued by the MDD are distributed to those who actually use the equipment (Barton, 1991). In a recent survey in the South West of England only 66 per cent of consultants and 33 per cent of junior anaesthetists were moderately confident that they had seen relevant ones (Weir and Wilson, 1991).

Organization. The hospital management must provide a safe environment for anaesthesia with well-designed and well-equipped anaesthetic, operating, and recovery rooms. The importance of skilled assistance — Operating Department Assistants or Anaesthetic Nurses, Recovery Nurses, and more senior anaesthetic help when required—has already been stressed in many mortality surveys (Buck *et al.*, 1987; Campling, 1992; Department of Health, 1991). A vexed issue of current importance is how much money should be spent on providing the monitoring that anaesthetists now consider necessary (Heath, 1988; Orkin, 1989). In an underfunded National Health Service, managers require proof that investment in safety is worthwhile. Unfortunately it is almost impossible to prove that a particular item of monitoring makes anaesthesia safer. A manager may give a low funding priority to improving monitoring standards because the risks from the use of current equipment is very low. However, Brahams (1989) points out that those who plead lack of cash may regret their decision when a case comes before the courts.

External pressure to change policies is required. The Association of Anaesthetists has published *Recommendations for standards of monitoring during anaesthesia and recovery* (Anon., 1988) but many of these are just recommendations rather than requirements. Inspectors from the Royal College of Anaesthetists visit hospitals to ensure that training is adequate. Ultimately, accreditation for training can be withdrawn if the facilities (including the standard of monitoring) are inadequate. The hospital manager would then lose a large section of the anaesthetic workforce — the trainees. This is a slow, cumbersome, and uncertain threat unsuited to dealing with urgent deficiencies in equipment. Surprisingly, no government body can enforce safety standards for the patients by demanding adequate anaesthetic and monitoring equipment. This is in contrast to the Department of Health and Safety which enforces safety measures to protect hospital workers. In many parts of the USA the driving force to improve standards has come from the insurance companies and in consequence minimal monitoring standards are mandatory.

An equally, if not more important, issue is the pressure to increase the efficiency in what is now commonly referred to as the 'Health Industry'. Pressure is exerted by hospital managers to increase the throughput of surgical patients while at the same time maintaining often already inadequate staffing levels and imposing other financial constraints. Consequently it becomes increasingly difficult to maintain safe standards of practice as working procedures are edged into the grey area between ideal and un-acceptable practice. Gaba (1987, 1989) warns that this creates a dangerous error-inducing system.

SAFE TECHNIQUES

There are obvious advantages in adopting techniques, whenever possible, that avoid known risks. For example, the complications of failed intubation (which includes death) will not arise if anaesthetic techniques are chosen that do not require tracheal intubation. Furthermore, it is wise to choose tech-niques that 'fail safe' (Hawkins, 1987). A spontaneously breathing patient will survive a disconnection whereas a paralysed patient may die.

MONITORING AND RECOVERY TECHNIQUES

Even if an error occurs, harm may be avoided if the incident is detected early and a recovery procedure implemented. Gaba *et al.* (1987) point out that

93 per cent of Cooper *et al.*'s (1984) critical incidents were successfully managed. Early detection allows more time to initiate recovery procedures and makes it less likely that an incident will produce harm to the patient (Schreiber and Schreiber, 1987; Gaba *et al.*, 1987). Factors affecting the recovery sequence have been analysed by Galletly and Mushet (1991).

How effectively monitoring improves safety is a controversial issue and arguments have raged over the benefits of pulse oximetry and 'mandatory minimal monitoring standards'. Hard evidence is lacking and, to all intents and purposes, may be impossible to obtain (Cooper *et al.*, 1987; Orkin, 1989; Keats, 1990; Duncan and Cohen, 1991).

Prompt corrective action may not allow much time for thought and some incidents may occur so rarely that their proper control may be hard to determine. Furthermore, it is vital that the recovery action taken is appropriate because an incorrect response may precipitate another incident (DeAnda and Gaba, 1990) which may have even more serious consequences. For these reasons there is much to be said for having readily available a set of Anaesthesia Action Plans (e.g. Eaton and Wilson, 1991). Such a strategy represents a shift to rule-based behaviour, rather than the slower knowledge-based behaviour. However, it may not be applicable to more unusual events (Reason, 1990). Decisions about emergency procedures should be made in advance and at leisure with the benefit of collective wisdom and then incorporated into the individual or department's standard operating procedures, thus increasing the amount of rule-based behaviour. Memory can be supported with a notebook containing drug dosages, infrequently required technical details, and emergency action plans (e.g. Eaton and Wilson, 1991). Such *aides-mémoire* should be used more commonly. DeAnda and Gaba (1990, 1991) and Schwid and O'Donnell (1992), using an anaesthesia simulator, have provided both a valuable training tool and also insight into the way anaesthetists deal with critical incidents. Proceeding further in this direction Gaba *et al.* (1991) have encouraged specific training in crisis management.

Conclusion

Despite continuing improvements in equipment design, accidents still occur and humans continue to make errors. This is inescapable and obvious, even in our everyday lives. But human errors in anaesthesia which may lead to death during a trivial surgical operation lead to understandable outrage. The public demands constant vigilance and perfect performance all the time, yet this is beyond the ability of human performance. Accidents and error will continue to occur despite every attempt to eliminate them. Therefore it is necessary to emphasize ways of detecting and recovering from them.

The current trend is to rely on sophisticated, expensive monitoring systems to detect incidents. Although there is no proof that they are effective, those of us who have become accustomed to using well-designed versions of this equipment have no doubts about their value.

The key to preventing an incident propagating and causing harm is the speedy employment of an effective recovery technique. Anaesthetists need to be well-trained in these procedures (DeAnda and Gaba, 1991; Schwid and O'Donnell, 1992). Novel problems may require new solutions and thus anaesthetists must also be trained to check the appropriateness of any rule, consider alternative possibilities, and be prepared to develop new solutions.

Many questions remain unresolved. What is the most effective monitoring and alarm system? What is the best strategy for choosing a recovery technique? How can we teach anaesthetists to deal with anaesthetic accidents? Why is there inertia against improving safety? How can a balance be struck between patient throughput and safety? Finally, what is safety worth?

References

Aitkenhead, A. R. (1991). Risk management in anaesthesia 1991. *Journal of the Medical Defence Union* **4**, 86–90.

Anon. (1988). *Recommendations for standards of monitoring during anaesthesia and recovery*. Association of Anaesthetists of Great Britain and Ireland, London.

Anon. (1992). *Stress and the medical profession*. British Medical Association, London.

Armstrong, J. N. and Davies, J. M. (1991). A systematic method for the investigation of anaesthetic incidents. *Canadian Journal of Anaesthesia* **38**, 1033–35.

Barton, A. (1991). 'Alarm signals' over warning signs? *Anaesthesia* **46**, 809.

Brahams, D. (1989). Anaesthesia and the law. Monitoring. *Anaesthesia* **44**, 606–607.

Buck, N., Devlin, H. B., and Lunn, J. N. (1987). *The report of the confidential enquiry into perioperative deaths*. Nuffield Provincial Hospitals Trust/King Edward's Hospitals Fund, London.

Campling, E. A., Devlin, H. B., Hoile, R. W., and Lunn, J. N. (1992). *The report of the national confidential enquiry into perioperative deaths 1990*. London.

Caplan, R. A., Posner, K., Ward, R. J., and Cheney, F. W. (1988). Peer reviewer agreement for major anesthetic mishaps. *Quality Review Bulletin* **14**, 363–68.

Caplan, R. A., Posner, K. L., Ward, R. J., and Cheney, F. W. (1990). Adverse respiratory events in anaesthesia: a closed claims analysis. *Anesthesiology* **72**, 828–33.

Caplan, R. A., Posner, K. L., and Cheney, F. W. (1991). Effect of outcome on physician judgements of appropriateness of care. *Journal of American Medical Association* **265**, 1957–60.

Cheney, P. W., Posner, K., Caplan, R. A., and Ward, R. J. (1989). Standard of care and anesthesia liability. *Journal of the American Medical Association* **261**, 1599–1603.

Chopra, V., Bovill, J. G., Spierdijk, J., and Koornneef, F. (1992). Reported significant observations during anaesthesia: a prospective analysis over an 18-month period. *British Journal of Anaesthesia* **68**, 13–17.

Cook, R. I., Woods, D. D., and McDonald, J. S. (1991). *Human performance in anesthesia* (CSEL 91.003). CSEL Report.

Cooper, J. B., Newbower, R. S., Long, C. D., and McPeek, B. (1978). Preventable anesthesia mishaps: a study of human factors. *Anesthesiology* **49**, 399–406.

Cooper, J. B., Long, C. D., Newbower, R. S., and Philip, J. H. (1982). Critical incidents associated with intraoperative exchanges of anesthesia personnel. *Anesthesiology* **56**, 456–61.

Cooper, J. B., Newbower, R. S., and Kitz, R. J. (1984). An analysis of major errors and equipment failures in anesthesia management: considerations for prevention and detection. *Anesthesiology* **60**, 34–42.

Cooper, J. B., Cullen, D. J., Nemaski, R., Hoaglan, D. C., Gevirtz, C. C., Csete, M., and Venable, C. *et al.* (1987). Effects of information feedback and pulse oximetry on the incidence of anesthesia complications. *Anesthesiology* **67**, 686–94.

Craig, J. and Wilson, M. E. (1981). A survey of anaesthetic misadventures. *Anaesthesia* **36**, 933–36.

Currie, M. (1989). A prospective survey of anaesthetic critical events in a teaching hospital. *Anaesthesia and Intensive Care* **17**, 403–11.

DeAnda, A. and Gaba, D. M. (1990). Unplanned incidents during comprehensive anesthesia simulation. *Anesthesia and Analgesia* **71**, 77–82.

DeAnda, A. and Gaba, D. M. (1991). Role of experience in the response to simulated critical incidents. *Anesthesia and Analgesia* **72**, 308–15.

Department of Health, Welsh Office, Scottish Home and Health Department, and DHSS, Northern Ireland (1991). *Report on confidential enquiries into maternal deaths in the United Kingdom 1985–1987*. HMSO, London.

Duncan, P. G. and Cohen, M. M. (1991). Pulse oximetry and capnography in anaesthetic practice: an epidemiological appraisal. *Canadian Journal of Anaesthesia* **38**, 619–25.

Eagle, C. J, Davies, J. M., and Reason, J. (1992). Accident analysis of large-scale technological disasters applied to an anaesthetic complication. *Canadian Journal of Anaesthesia* **39**, 118–22.

Eaton, J. M., and Wilson, M. E. (1991). *Anaesthesia action Plans*. Abbott Laboratories Ltd, Maidenhead.

Gaba, D. M. (1989). Human error in anesthetic mishaps. *International anesthesiology clinics* **27**, 137–47.

Gaba, D. M. and Lee, T. (1991). Measuring the workload of the anesthesiologist. *Anesthesia and Analgesia* **71**, 354–61.

Gaba, D. M., Maxwell, M., and DeAnda, A. (1987). Anesthetic mishaps: breaking the chain of accident evolution. *Anesthesiology* **66**, 670–76.

Gaba, D. M., Howard, S. K., Fish, K. J., Yang, G., and Sarnquist, F. H. (1991). Anesthesia crisis resource management training. *Anesthesiology* **75**, A1062.

Galletly, D. C. and Mushet, N. N. (1991). Anaesthesia system errors. *Anaesthesia and Intensive Care* **19**, 66–73.

Gannon, K. (1991). Mortality associated with anaesthesia. A case review study. *Anaesthesia* **46**, 962–66.

Gravenstein, J. S., Cooper, J. B., and Orkin, F. K. (1990). Work and rest cycles in anesthesia practice. *Anesthesiology* **72**, 737–42.

Hawkins, F. H. (1987). *Human factors in flight*. Gower Technical Press, Aldershot, Hants.

Heath, M. L. (1988). The cost of safety. In *Some aspects of anesthetic safety* (ed. O. P. Dinnick and P. W. Thompson), pp. 401–18. Baillière's Clinical Anaesthesiology. Baillière Tindall, London.

Keats, A. S. (1990). Anesthetic mortality in perspective. *Anesthesia and Analgesia* **71**, 113–19.

Kumar, V., Barcellos, W. A., Mehta, M. P., and Carter, J. G. (1988). An analysis of critical incidents in a teaching department for quality assurance. A survey of mishaps during anaesthesia. *Anaesthesia* **43**, 879–83.

McIntyre, J. W. R. (1991). Alarms in the operating room. *Canadian Journal of Anaesthesia* **38**, 951–53.

McKay, W. P. S. and Noble, W. H. (1988). Critical incidents detected by pulse oximetry during anaesthesia. *Canadian Journal of Anaesthesia* **35**, 265–69.

March, M. G. and Crowley, J. J. (1991). An evaluation of anesthesiologist's present checkout methods and the validity of the FDA checklist. *Anesthesiology* **75**, 724–29.

Mayor, A. H. and Eaton, J. M. (1992). Anaesthetic machine checking practices: a survey. *Anaesthesia* **47**, 866–68.

Newbower, R. S., Cooper, J. B., and Long, C. D. (1980). Failure analysis — the human element. In *Essential noninvasive monitoring in anesthesia*. ed. J. S. Gravenstein, R. S. Newbower, A. K. Ream, *et al.*, pp 269–82. Grune and Stratton, New York.

Orkin, F. K. (1989). Practice standards: the Midas touch or the Emperor's new clothes. *Anesthesiology* **70**, 567–71.

Posner, K. L., Caplan, R. A., and Cheney, F. W. (1991). Physician agreement in judging clinical performance. *Anesthesiology* **75**, A1058.

Reason, J. (1990). *Human error*. Cambridge University Press, New York.

Schreiber, P. and Schreiber, J. (1987). *Anesthesia system risk analysis and risk reduction*. North America Drager, Telford, Pennsylvania.

Schwid, H. A. and O'Donnell, D. (1992). Anesthesiologist's management of simulated critical incidents. *Anesthesiology* **76**, 495–501.

Thompson, P. W. (1987). Safer design of anaesthetic machines. *British Journal of Anaesthesia* **59**, 913.

Weinger, M. B. and Englund, C. E. (1990). Ergonomic and human factors affecting anesthetic vigilance and monitoring performance in the operating room environment. *Anesthesiology* **73**, 995–1021.

Weir, P. M. and Wilson, M. E. (1991). Are you getting the message? A look at the communication between the Department of Health, manufacturers and anaesthetists. *Anaesthesia* **46**, 845–48.

Wilson, M. E. (1985). Morbidity and mortality in anaesthetic practice. In *Recent advances in anaesthesia and analgesia*. (ed. R. S. Atkinson, and A. P. Adams), Vol. 15 pp. 209–23. Longman, London.

Wilson, M. E. (1988). To err is human ..... *Health Equipment and Supplies* (Sept.), 58.

6

Selecting and educating safer doctors

CHRIS McMANUS and CHARLES VINCENT

Reducing medical accidents by better selection of medical students or medical specialists is an immensely attractive idea; more rigorous, better targeted selection procedures might, at one fell swoop, have a long-term impact in increasing safety, and raising overall standards of medical care. This at least is a frequently used argument. However, such proposals invite a host of difficult questions. Does selection mean searching for doctors of higher general competence, or is it principally seeking to weed out a few 'bad apples' who are implicitly assumed to be responsible for most medical accidents? Is involvement in medical accidents, or any other kind of accident, a stable individual characteristic? In other words, are some individuals naturally careless or accident prone? Even if such people can be identified, can they be identified at seventeen years of age, prior to medical school entry? And if they can, is it justifiable to reject such an applicant solely on those grounds when they might later change their attitudes and behaviour, and become competent or even excellent doctors?

There is little empirical research to either support or reject the view that safer doctors can be selected, although there is relevant work in related areas. In this chapter we shall argue that attempts to alter medical selection are premature until the underlying assumptions have been thoroughly examined. In particular the feasibility of selection as a policy instrument requires careful investigation, together with the short- and long-term effects and the theoretical and practical limitations. We argue that although selection procedures in medicine show great scope for improvement, selection for safety *per se* should not be a priority. Instead, attention should be concentrated on those defining characteristics which allow the training of a competent doctor; and for post-graduate selection these will vary by specialty. At the undergraduate level they should be for what we call *canonical characteristics* — generic abilities which provide an adequate substrate for medical education, general training, and continuing professional development. Identification, measurement, and validation of the characteristics which correlate with competency will thereby allow the development of selection procedures founded on a secure theoretical and empirical basis.

This chapter will firstly ask whether the concept of accident-proneness as a stable personality characteristic is supported by research data, and whether other personality measures are predictive of accident liability. We will then present a theoretical argument to identify the limits of successful selection and compare the advantages to be gained from selection with those that may be achieved by training. These arguments will be illustrated with hypothetical examples of different selection strategies for entry to medical school. The same basic arguments, however, apply in respect of selection for a particular specialty or at any career point.

The concept of accident-proneness?

The reduction of accidents by improved selection requires that the likelihood of having accidents is a relatively enduring individual characteristic. If such a tendency is not fairly stable then selection at age eighteen will have little influence on behaviour twenty or thirty years later. Early studies searched principally for an 'accident-prone personality' which would influence accidents throughout a lifetime.

The concept of accident-proneness dates back to the studies of Greenwood *et al.* (1919), Greenwood and Yule (1920), and Newbold (1926) in factory accidents, who showed that workers' individual accident rates were not a random, Poisson distribution (Fisher, 1973), but were better fitted by a negative binomial model with individuals differing in their propensity for accidents. A recent methodological advance emphasizes that the simple use of accident rates is statistically insensitive to individual differences, and that more power is gained by examining the distribution of time-intervals between events (i.e. accidents); it then seems incontrovertible that individuals differ in their accident liability (Shaw and Sichel, 1971), although this need not be explained in terms of accident-proneness. Alternative explanations are that individuals are, for some reason, differentially exposed to risk and that accidents themselves can alter the risk of future accidents.

The problems implicit in using the concept of accident-proneness as a tool for reducing accidents are seen in the large study of Cobb who studied 29 531 Connecticut drivers during 1931–33 and 1934–46. The results suggested that 4 per cent of drivers caused 36 per cent of the road accidents (Cobb, 1939). Further analysis showed that, compared with those with no accidents in the same period, drivers having four accidents in 1931–33 were nearly seven times more likely to have accidents during 1934–46. Cobb concluded: 'once a group has been established as being predominantly accident-free or accident-liable, its future history as a group can be predicted with astonishing accuracy from its past performance'. However, Forbes (1939) emphasized that only 1.3 per cent of drivers could thus be called 'accident-prone', and that their removal on the basis of accidents during 1931–33 would have decreased accidents during 1934–36 by only 3.7 per cent. As we shall see it is also doubtful, to say the least, that the 1.3 per cent of accident-prone drivers could be identified prior to their involvement in an accident.

The problem with studies such as those of Cobb is their confusion of two separate epidemiological concepts. *Relative risk* assesses how a factor alters the likelihood of an outcome for an individual (e.g. in hypertensives, smoking doubles the cardiovascular mortality), whereas *attributable risk* assesses the population risk of a condition due to a factor (e.g. in a group of hypertensives, there will be seven excess deaths due to smoking per 1000 patient years). As has been emphasized (Rose and Barker, 1986), it is attributable risk and not relative risk which should be the guide to management and policy decisions. Here then is the crunch for using accident-proneness as a tool for selection; although the relative risk of some individuals having accidents may be really quite high compared with other individuals, the attributable risk due to these individuals is only a small proportion of the total number of accidents. To put it another way, the sensitivity of the test may be high, but its specificity may be low.

McKenna (1983) has argued that the concept of accident-proneness is conceptually confused and should be replaced by 'differential accident involvement', a term that '...does not prejudge the issue.... It is an area of study not a set theory'. Differential accident involvement accepts that some individuals may be more liable to error or accident at certain periods, perhaps in response to life-events, depression, or mood-shifts (Irwin, 1964); and it would indeed seem that life-events (Selzer and Vinokur, 1974; Alkov and Borowsky, 1980; Stuart and Brown, 1981; Whitlock *et al.*, 1977), and mood (at least as indicated by prior suicidal intentions (Kaplan and Pokorny, 1976)) do relate to accidents. The duration of such periods of accident vulnerability, their relation to accidents in subsequent periods, and the personality correlates which might eventually allow the identification of a syndrome of accident-proneness, are left open. Instead, emphasis is placed on understanding the mechanisms that might underlie differential accident involvement, and McKenna suggests that such a knowledge of the psychological processes underlying human error would encourage training where appropriate, selection when necessary, and redesign of equipment or working environment where possible.

Personal characteristics predisposing to accidents

Although medical accidents are little studied, road traffic accidents (RTAs) have been extensively investigated and may act as a provisional model for some kinds of medical accidents. In this section we will draw heavily upon the reviews and research of West and his colleagues (French *et al.*, 1992; West *et al.*, 1992a, 1992b).

RTAs have been related to several personality measures, including extraversion (Pestonjee and Singh, 1980; Fine, 1963), sensation-seeking (Loo, 1979), neuroticism (Shaw and Sichel, 1971), type A behaviour (Perry, 1986; Evans *et al.*, 1987), and risk-taking behaviours (Jonah, 1986), although not all studies have supported these findings (Wilson and Greensmith, 1983; Singh, 1978; Craske, 1968). Additionally it has been suggested that accidents are associated with increased aggression, seeking of prestige, and competitiveness (McGuire, 1972). There is therefore some evidence for differential accident involvement, although it is not clear how such characteristics might manifest in a medical environment. Decision-making strategies, however, have an obvious relevance to clinical medicine.

Decision-making style and accidents. West has argued that since many accidents involve errors of decision-making, then individuals may vary in their characteristic decision-making style, which in turn may predict accident rates in particular situations. Similarly Jensen (1982) has emphasized that 80–85 per cent of flying accidents are attributed to 'pilot error', and that probably 50 per cent of these are due to errors of judgement. West has developed the Decision Making Questionnaire (DMQ) (West *et al.*, 1992b; French *et al.*, 1992) which has 21 questions representing seven independent factorial dimensions, called Thoroughness, Control, Hesitancy, Social Resistance, Perfectionism, Idealism, and Instinctiveness. The dimension of Thoroughness (deliberate and logical decision-making, planning well ahead, working out pros and cons) was inversely correlated with six different measures of driving style, in particular with a tendency to excessive speed, and hence with frequency of accidents.

Instinctiveness (relying on 'gut feelings', sticking by decisions come what may) also correlated with accident frequency, although not with other driving behaviours. Other dimensions also related to driving behaviour, but not to accidents *per se*. West *et al.* (1992a) found that thoroughness was a significant prospective predictor of accident rates over a two-year period. The psychological nature of thoroughness is not clear, but it might reflect a more global trait such as impatience, particularly since type A behaviour, extraversion and sensation-seeking, which all correlate with accident involvement, all contain components which could be described as impatience (French *et al.*, 1992).

It is worth noting that in the study of West *et al.* (1992b) the two–year test–retest correlation of thoroughness is 0.46, and that the correlation of accident rates in one period with the next is only 0.08. Temporal stability in the measured characteristics is not therefore high. The result is that the correlation between thoroughness and accident rate cannot be high, and is in fact only –0.18. However, disattenuation of the correlation between accident-rate and thoroughness, taking the low reliability of each measure into account, gives a correlation of −0.938. Thoroughness might therefore be a strong predictor of accident involvement if both could be measured reliably. The implication is that the measures are tapping an important causal origin of some relatively constant component of accident proneness, although the majority of variance in number of accidents is probably due to other factors (see Chapter 1).

Social attitudes and accidents. Road traffic accidents have frequently been associated with a variety of variables which are broadly encompassed under the heading of 'mild social deviance' (West *et al.*, 1992a); thus links have been found with 'expression of hostile impulses' (Conger *et al.*, 1959; Barmack and Payne, 1961; Harano *et al.*, 1975; Schuman *et al.*, 1967; Hertz, 1970), 'eccentricity, impulsivity, or mild psychopathy' (McFarland, 1968), and 'social deviance' (Suchman, 1970). West *et al.* (1992a) developed a 10-item questionnaire, the 'Social Motivation Questionnaire', which assessed the extent to which individuals would take part in behaviours that could be construed as minor social deviance (e.g. parking on double yellow lines, travelling on public transport without paying a fare, or not declaring cash payments to the Inland Revenue). The measure was correlated with driving speed and deviant driving behaviour, and also with low thoroughness on the DMQ. Additionally, it was correlated with accident rates, and only part of that relationship was explained by higher driving speeds. The psychological substrate of mild social deviance is not clear, but West *et al.* (1992a) suggest that it probably reflects 'greater emphasis on the need to make good progress with less consideration of the adverse consequences of an accident'. It should also be noted that mild social deviance was less in females than males.

Although there are a few members of any profession who breech its rules and so could be described as deviant, it might seem that 'mild social deviance' has little relevance in a medical context. However, acting on impulse without sufficient regard for the consequences might well lead to accidents. There is certainly evidence that the confidence of some junior doctors far exceeds their abilities in some areas (see Chapter 7), and an attitude of 'medical machismo' which leads to a determination to handle any emergency oneself (whatever the costs to the patient) has been documented for decades by medical sociologists (see, for example, Merton *et al.*, 1957; Coombs and Stein, 1971; Becker *et al.*, 1961; Bloom, 1973). Where a junior doctor feels that calls for assistance might be regarded as evidence of weakness or incompetence, this tendency is exacerbated still further.

Characteristics of doctors involved in medical accidents. There are few studies which have related the likelihood of medical accidents to stable individual characteristics of doctors. The sole exceptions, which are of some importance, are two recent studies of malpractice experience (Sloan *et al.*, 1989; Kravitz *et al.*, 1991). These studies examined the qualifications and training of doctors involved in accidents, hypothesizing that such factors might relate to accident involvement. However, doctors with more prestigious credentials, who were board-certified, who were from the top third of US medical schools, or had degrees from North American universities were no less likely to have malpractice claims than other doctors (Sloan *et al.*, 1991); neither did experience with medical research or teaching relate to rate of malpractice claims. 'The general conclusion is that past claims experience is only modestly predictive of intrinsic claims proneness. Although physicians incurring large numbers of claims in the past are more likely, on the average, to incur large numbers in the future, predictions about individuals based on past claims experience are probably not accurate enough to identify most claims-prone physicians or to allow reliable judgements about an individual's propensity to practice negligently in the future' (Rolph *et al.*, 1991).

In both studies, however, female doctors had lower claim rates than males (Sloan *et al.*, 1991; Rolph *et al.*, 1991). Women, it should be noted, also have lower rates of road traffic accidents (Evans, 1991; Maycock *et al.*, 1991). The reasons for sex differences in behaviour are complex (Maccoby and Jacklin, 1975; Halpern, 1992) and not yet fully understood. Nevertheless it is possible that the difference in malpractice claim rates (and by implication, medical accidents) relates to an underlying personality characteristic that is stronger in men — impulsivity would seem to be a strong candidate. However, there is only a small overlap between accidents and negligence claims; most cases of negligence do not result in claims and many claims are filed in the absence of any negligence (see Chapter 2). It may therefore be not so much that women are less liable to be involved in accidents but that they are less likely to be sued. This is broadly the conclusion of Sloan *et al.* (1989) who suggest that the result 'reflects a patient-physician practice style [by women] that is less conducive to claims' (Sloan *et al.*, 1991).

To summarize, certain personality traits and decision-making styles, perhaps reflecting an underlying dimension of impatience/thoroughness, might predispose to accident involvement. Other dimensions may be revealed in subsequent research. In medicine we know only that female doctors are less likely to be sued, which may mean that they are involved in fewer accidents; this may be because they are more thorough, or it may reflect differences in communication style with both colleagues and patients. The next section examines whether, if stable predictors of accidents could be identified, it would be feasible to select for them.

The selection process

There is an extensive literature on selection procedures, dating from the early years of the century. After a period of pessimism and decline in the 1950s and 1960s, interest has grown steadily both in industry and in academic circles, and the usefulness and validity of some selection procedures has been established (see, for example, Schmidt *et al.*, 1992). A great variety of selection procedures has been developed, some of the main ones being: interviews (both structured and unstructured); interviews in which candidates are asked to predict their behaviour in certain situations; tests of intellectual ability, perceptual-motor skills, personality and attitudes; tests which simulate or involve the work to be done; computer-assisted tests; taking up references; graphology; and peer assessment (Robertson and Smith, 1989).

In establishing the validity of any selection procedure it is first necessary to define outcome measures, the desirable or essential skills or abilities which are assessed when selection procedures are evaluated. In medicine this might involve assessing clinical knowledge, diagnostic skills, technical competence, ability to communicate with staff and patients and any other characteristic that would enhance clinical standards and, in our case, promote safety. The validity of selection is established by comparing the results of the selection with candidates' later scores on the outcome measures. Selection procedures with the best predictive accuracy are work sample tests and tests of general intellectual ability and, where applicable, psychomotor ability. Supervisor/ peer assessments, assessments centres, biographical data, and general mental ability are the best remaining methods. References, interviews, personality assessment, and interest inventories are very poor predictors (Robertson and Smith, 1989). Selection for medical school is initially based on examination results, probably a reasonable reflection of general mental ability and thus a useful part of the selection process. In the UK results in A-level examinations are the best predictors of success during selection (McManus and Richards, 1984; McManus *et al.*, 1989b), and they are also to some extent predictive of success later during the medical course (McManus and Richards, 1986). Medical student selection also relies extensively on interviews and references, which are among the least valid of all selection procedures (Robertson and Smith, 1989); nevertheless, evidence suggests that medical school shortlisters and interviewers can at least be reliable in the judgements that they make (Richards *et al.*, 1988; McManus *et al.*, 1989a; McManus and Richards, 1989).

Selecting for safety. Our comparatively positive view of the use and validity of selection procedures does not extend into the area of safety, although one might think that this would be a prime concern in industry. In a survey of research on outcome measures in selection, Landy and Rastegary (1989) found that only four studies out of 408 reviewed examined accident rates. Their comments on this state of affairs are worth quoting in full:

The simple fact is that virtually no one is studying accidents from the perspective of individual differences among incumbents. There is certainly a great deal of discussion about safe behaviour in environments such as nuclear power plants, air traffic control towers and airplane cockpits. Nevertheless, the published literature on these topics from a selection procedure is non-existent. Needless to say, the number of industrial deaths and lost time injuries remains unacceptably high. It is hard to believe that applied psychologists have run out of ways to study or understand safety behaviour. It may be that the differential psychologist has simply deferred to the human factors psychologist to solve the problem. This may be a little premature. At best the answer to safe behaviour is likely to come from the joint efforts of the personnel psychologist and the human factors psychologist rather than from the unique contribution of either of them.

We can illustrate this lamentable state of affairs by briefly commenting on the selection of pilots. The selection of pilots has always attracted great attention, due to the high costs of training and the serious consequences of accidents. Typically the outcome measures for validation are performance in training or post-training probationary flying, rather than a specific assessment of errors or of unsafe behaviour. Although a range of pencil-and-paper and computer-based tests have been used, the utility or validity of these is often assumed rather than proven. Detailed studies suggest that validity is often present only when the tests are close in time to outcome measures, and/or share their characteristics (as for instance with a computerized aircraft landing task for selecting trainee pilots (Fowler, 1981); as Bale *et al.* (1973) suggest, it is difficult or impossible to find tests which generalize across a broad range of validation criteria. The result is that selection measures are almost unrelated to outcome measures at the end of training (Bale *et al.*, 1973). Occasionally psychometric tests do show differences between successes and dropouts during training, although the sensitivity is poor. The result is that failures can only be eliminated at the selection stage by rejecting large numbers of probable successes.

Given that selection procedures in other fields have been relatively successful, we do not wish to suggest that selection for safety is a lost cause. Landy and Rastegary (1989) argue that, although research on selection in relation to safety is non-existent, the attempt to select for safety should certainly be made. What benefits might we gain from attempting to select safer doctors? Is the project worthwhile, or could the effort involved be better directed at some other accident-reducing endeavour? The next sections consider these fundamental questions in some detail.

Medical student selection and training: a theoretical model

The limits of selection. Student selection illustrates well the subtleties of a seemingly straightforward process. In essence, selection is extremely simple. A number of students, N, applies to study medicine, and a smaller number M is

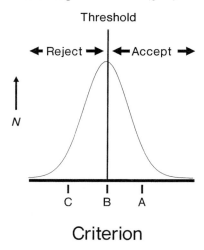

Criterion

Figure 6.1 Selection on a single criterion. Individuals differ in their scores on the criterion, with individual A scoring high, B average, and C scoring low. Selection involves the setting of a threshold, those above the threshold being selected and those below being rejected; thus A is selected, C is rejected and B is marginal.

accepted as medical students; the figure N/M is known as the *selection ratio*. Although, in the UK at least, the selection ratio for individual medical schools is high (between 5 and 20), the selection ratio for the system as a whole is much lower because each candidate applies to four or five medical schools. In fact there is overall approximately one reject for every student selected and we will assume in our discussion that the selection ratio is 2. This selection ratio is far lower than is generally thought to be these case; thus it is *not* the case that 'medical schools have been blessed with an army of applicants' (Anon., 1992).

Consider a system using *single criterion selection* (usually academic or intellectual ability, although other measures might be used). Typically, scores will be normally distributed in the population; if 50 per cent of candidates are to be selected then selection will simply involve choosing those half of the applicants with the highest scores (see Fig. 6.1).

If selection is on the basis of two criteria (such as 'academic ability' and 'ability to communicate') then the situation is more complex. Assume that the two abilities are uncorrelated (a condition that can always be met by considering their principal components). In Fig. 6.2(a) a set of individual candidates, assumed distributed as a bivariate normal, are shown as points on the graph. Selection can take several forms. Figure 6.2(b) shows *inclusive selection*, in which candidates are selected only if they score above a certain level on *both* criteria, whereas Fig. 6.2(c) shows *exclusive selection* in which

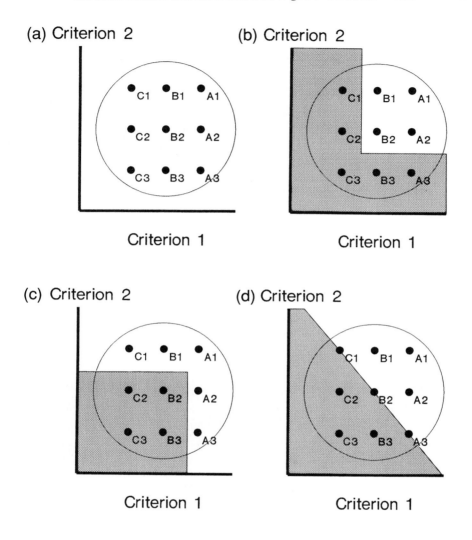

Figure 6.2 Selection on two independent criteria, each scored on one of the two axes. Individual candidates are shown as points: A1, A2, and A3 score highly on criterion 1, A1, B1, and C1 score highly on criterion 2; candidate A1 scores highly on both criteria, and candidate C3 scores poorly on both criteria. In inclusive selection (b), candidates are selected if they score highly on criterion 1 *and* criterion 2; the individuals in the shaded area are therefore rejected. In exclusive selection (c), candidates are selected if they score highly on criterion 1 *or* criterion 2. In compensatory selection a relatively poor performance on one criterion is compensated by better performance on the other criterion.

candidates are selected who score above a certain level on *either* criterion; finally, Fig. 6.2d shows *compensatory selection* in which a relatively high score on one criterion offsets a relatively low score on the other criterion. There are, of course, also selection strategies which can be seen as combinations of the methods just described.

The addition of a second selection criterion has important effects which are different for each of the selection methods.

1. *Inclusive selection.* The minimum standard set on *each* criterion (Fig. 6.2(b)) has to be *lower* than if only a single criterion had been used; thus candidate B in Fig. 6.1 is borderline, whereas candidate B2 in Fig. 6.2(b) is safely admitted, despite each candidate being average on each criterion. Inclusive criteria therefore reduce the average standards on each criterion, but with the advantage of ensuring that all entrants satisfy all of a range of minimum standards.

2. *Exclusive selection,* in contrast, means that *some* entrants score very highly on each criterion but at the expense of generally scoring very poorly on the other criterion; thus candidate B2 would be rejected as not being outstanding on *either* criterion, whereas candidate C1 is accepted because of their high score on criterion 2, despite having a score on criterion 1 which would have led to outright rejection under the single criterion selection process of Fig. 6.1. Exclusive criteria therefore result in high variance between candidates, who always perform at a high level on at least one criterion, but typically perform poorly on the other criterion.

3. *Compensatory selection* is a compromise between inclusive and exclusive selection. Candidates who perform *very* poorly on one criterion are only selected if they also perform *very* well on the other criterion. The method still accepts some entrants who are particularly poor on one or other criterion; thus candidate C1 is now borderline, whereas under single-criterion selection they would have been rejected; and candidate A3 is also borderline despite being clearly accepted under single-criterion selection.

All three methods for using two selection criteria have a single feature which is necessarily common to all; *the average performance of entrants on any one of the criteria will be less than if that criterion had been the sole basis for selection.* To put it more concretely; if one wishes to select candidates who are not only academically able but are also selected for their communicative ability, empathy, dexterity, or other possible correlates of lower rates of medical accidents, then that is only possible by selecting candidates with lower overall levels of academic ability than would have been achieved if academic ability were the sole criterion for selection. Exactly the

Table 6.1   The effects of selection on multiple criteria upon selection on one of those criteria. Using a selection ratio of 2, the table shows the proportion of candidates selected on one of the criteria (1) as the number of criteria increases.

Number of selection criteria	Proportion entrants rejected on criterion 1
1	Bottom 50%
2	Bottom 29.3%
3	Bottom 20.6%
4	Bottom 15.9%
5	Bottom 12.9%
6	Bottom 10.9%
10	Bottom 6.7%
20	Bottom 3.4%
50	Bottom 1.4%

same arguments apply when candidates are being considered for post-graduate training in surgery, general medicine, psychiatry, or other specialties. The use of multiple criteria might therefore be counter-productive.

Multiple selection criteria. Lists of the desirable characteristics that potential doctors should have are often lengthy. Using multiple criteria during selection exacerbates the effects found with only two criteria. Consider the case of inclusive selection, where all candidates meet minimum standards on all criteria. If 50 per cent of candidates can be selected overall, then the top 70.7 per cent must be selected on criterion 1 *and* on criterion 2 (Fig. 6.2(b)). Only the bottom 29.3 per cent of candidates on each criterion are rejected, compared with 50 per cent with only one criterion. With three criteria only the bottom 20.6 per cent on criterion 1 will be rejected; and so on, as shown in Table 6.1. As the number of criteria increases so the effect is to exclude a progressively smaller and smaller number on each criterion. The consequence is that with the low selection ratio that is found in medical student selection, the use of multiple selection criteria cannot result in entrants who are *well* qualified on *all* criteria but instead only results in the rejection of candidates who score *particularly badly* on *at least one* criterion.

Selection and performance. If selection is to reduce medical accidents by selecting individuals with lower scores on a personality or other measure then individuals with that personality should have lower accident rates. We now consider a hypothetical example to illustrate the problems inherent in attempting to select for a particular outcome — in this case a reduced accident rate.

Figure 6.3(b) shows the distribution of some accident-related personality measure in the population, and Fig. 6.3(a) shows that the likelihood of an accident (taken arbitrarily over some time period) is three times greater in those with high proneness who are 2 standard deviations above the mean

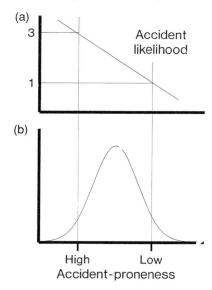

Figure 6.3 Variation in accident-proneness and its relation to the likelihood of accidents (hypothetical data). In (b) is shown the distribution of accident-proneness in the population, the dashed vertical lines indicating individuals on the 2.5th and 97.5th percentiles. In (a) is shown the relative likelihood of an accident for individuals of different proneness.

than in those with low proneness who are 2 standard deviations below the mean. One high-risk individual therefore causes as many accidents as three low-risk individuals. Nevertheless, because 95 per cent of individuals are between these extremes, the effect of selection upon the total accident rate is low. Using the numerical values quoted, the bottom 10 per cent of individuals cause only 15 per cent of accidents; and hence if those individuals were excluded from practising and replaced with others with a lower rate of accidents, then 94 per cent of accidents would still occur. Similarly, excluding from practice the bottom 20, 30, 40, or 50 per cent of individuals on this scale would reduce the number of accidents to only 90, 84, 81 and 78 per cent of their initial rate.

The potential effects of selection upon eventual accident rates are therefore quite small. To obtain even a 22 per cent reduction in accidents would require 50 per cent of applicants to be excluded (a figure equal to the selection ratio, and thereby implying selection on this criterion alone). This result means that selection would not be occurring on any other criterion (such as academic ability) — and that, of course, might result in *higher* rates of accidents for other reasons.

The conclusion seems to be that any realistic personality-type measure which could be used at the time of selection for differentiating applicants with a high risk of accidents would have only a small effect on the overall accident rate, and that to have a larger effect it would need to preclude any other form of selection, which might itself adversely affect the accident rate. This does not mean that selection is unimportant in relation to safety, only that safety *per se* should not dictate the selection criteria. Of course a concern with safety could and should be incorporated as a part of training. The effects of training must be considered in the next section.

The effects of training. Medical students enter medical school with almost zero ability to practice medicine (excluding a few rudimentary skills at first aid). Students will differ in their rate of acquisition of appropriate knowledge and skills, so that by the end of their five years of study they will differ in their ability to practice medicine competently. A small number, perhaps 10 per cent or so overall, will have insufficient knowledge and skills to pass exams at the end of various courses and will leave the school; the majority, however, will qualify. Figure 6.4 models the rate of growth of knowledge in individual students, two of whom fail at various stages of their course.

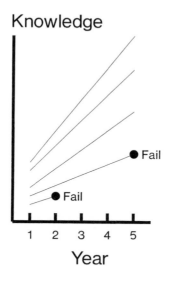

Figure 6.4 A model of the rate of growth of the absolute amount of medical knowledge in relation to time in the medical school. Two candidates acquire knowledge at a sufficiently low rate to mean that they are required to leave the medical school, either at the end of the second year or the fifth year. Amongst the three students who qualify there is a variation in the amounts of knowledge attained.

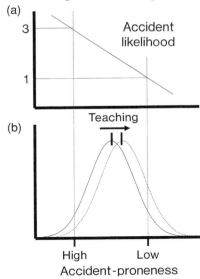

Figure 6.5 As in Fig. 6.3, individuals differ in their accident-proneness. The effect of training is to shift the distribution of proneness to the right, from the dotted to the solid line. The result is to reduce the overall likelihood of an accident.

The effect of good training and education is to encourage students to acquire more knowledge and, in particular, more *appropriate* knowledge, which is useful and applicable. We will assume that students who acquire appropriate knowledge are therefore less likely, on aggregate, to have accidents. The process can be modelled by assuming, as in Fig. 6.3, that individuals two standard deviations above and below the mean level of knowledge differ in their likelihood of an accident by a factor of three. The effect of training (Fig. 6.5(b)) is to increase the amount of knowledge and thereby to shift the distribution of accident proneness to the right, thereby making accidents less likely. If the shift to the right is by the fairly small amount of half a standard deviation (i.e. an individual on the 5th percentile acquires the knowledge of an individual on the 13th percentile) then the rate of accidents overall decreases to 75 per cent of its previous level; and if the shift is by one standard deviation, in which the individual on the 5th percentile achieves the knowledge level of an individual on the 26th percentile, then the accident rate falls to 57 per cent of its previous level.

Training even at relatively reasonable levels can therefore have effects upon the accident rate which can only be achieved through selection by concentrating selection upon a single criterion, and which would be difficult to achieve due to the scarcity of validated criteria selection measures. Training also has other advantages: it can be focused on the specific problem areas where accidents occur, so that skills, knowledge, and attitudes can be provided which are directly valid for the particular end-point. Selection can still proceed using other criteria that are not directly safety related. However, selection and training may interact; we may be able to select those doctors who are willing to learn and to practice safely.

Selection for training. Successful training requires an interaction between trainer and trainee. Motivation, learning experience, and study skills will all affect the efficiency and effectiveness of training, and to a large extent these attributes will be independent of simple intellectual or academic ability. Thus although medical school entrants may enter medical school with a *tabula rasa* for specific medical knowledge, they are not identical in ability to acquire that knowledge.

Figure 6 schematizes a situation in which entrants differ in ability to acquire knowledge and skills through training. In the first year this results in some differences in their eventual knowledge and ability. However, since understanding of a discipline provides a springboard for further learning, helps the integration of knowledge between different areas, and helps to motivate trainees for further learning then the greater gains of some students

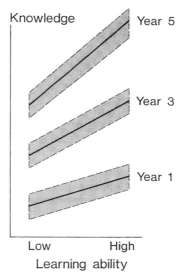

Figure 6.6   The absolute amount of medical knowledge acquired in individuals in their first, third, and fifth year at medical school. In each year there is variability, indicated by the shaded area, but those individuals with higher learning ability tend to acquire more knowledge than those with lower learning ability. These differences become accentuated across the years as later training becomes contingent upon more successful earlier training.

at the end of year 1 will be compounded during year 2, and so on. The result is that by the end of the course there can be relatively large differences between students.

The effects of selection are two-fold. Firstly, by selecting entrants who are in the upper part of the distribution for learning ability, the overall knowledge and ability of the students is increased. Secondly, the variability between students will be reduced, thereby helping to promote a 'culture' or 'ethos' in which independent, self-motivated learning is the norm, so that new knowledge continues to be acquired throughout post-graduate training. Selection for ability to be trained therefore has a 'multiplier effect', over and above that of simple personality differences, and makes it a particularly powerful form of selection.

What are the canonical characteristics to be found by selection?

We have argued that selection for specific, safety-related personality characteristics is unlikely to be an effective way of reducing accidents, and may indeed be counterproductive by reducing the average level of entrants on other criteria, such as academic ability. In contrast, training can have large effects on the likelihood of accidents, and can be progressively adapted and updated for the specific needs of particular students. If selection does have a role then it is likely to be principally in terms of selecting for 'ability to be trained', rather than in terms of personality measures. If that is the case then what are the *canonical characteristics* on which selection should be based? Here we are searching for a small number of broadly uncorrelated characteristics that can be measured, and which predict independent aspects of the ability to be trained at medical school, and to perform as a medical student and doctor.

At present the most important criterion used during student selection is academic ability (as measured by GCSE and A-level results in the UK). That is a sensible criterion, probably reflecting both intellectual ability and motivation to learn. Medical students have to learn prodigious quantities of information and must study for frequent exams, so that previous ability to study and pass exams is likely to be a useful predictor of success. Achievement at academic examinations is also correlated with general intellectual ability and there seems little doubt that IQ is a good measure of ability to be trained and to respond to diverse problems and challenges (McManus *et al.*, 1990).

Of course that is not to say that academic ability and intelligence should be the only criteria of selection. Although A-level results do correlate with success in university-level training, the correlation is not high, either in general (Bagg, 1970; Entwistle and Wilson, 1977; Choppin, 1979), or specifically in medicine (Savage, 1972; Mawhinney, 1976; Tomlinson *et al.*,

1977; Richardson, 1980; Lipton *et al.*, 1988). In part that reflects the reduced correlation that results from 'restriction of range' (Ghiselli *et al.*, 1981), but it also indicates that other measures are important. Studies of achievement in university students in general suggests that approaches to study are good predictors of success, and are independent of previous achievement (Entwistle *et al.*, 1979; Entwistle and Waterston, 1988; Biggs, 1978, 1985; Newble and Gordon, 1985). In particular, 'deep learning' seems to be a good correlate of achievement in academic disciplines, and 'strategic learning' of achievement in applied professions. 'Surface learning' in all subjects seems to be predictive of failure, and it is an approach which often seems to be encouraged by the type of curriculum found in medical schools (Coles, 1985; Tooth *et al.*, 1989). Learning ability, and in particular deep or strategic learning, are therefore good candidates as selection criteria. Much selection uses biographical information about individuals (biodata) (Asher, 1972), and in general such measures show adequate reliability and validity (Herriot, 1984; Asher, 1972; Owens *et al.*, 1966; Owens, 1976; Reilly and Chao, 1982; Asher and Sciarrino, 1974). In so far as they may also be indirect indicators of learning style they could also be useful selection criteria in medicine. In particular, extensive achievement in hobbies, be they musical, sporting or other, indicates self-motivation and an ability to direct one's study which probably correlates with strategic learning. And similarly, interest in and participation in community activities, particularly those involved with medical or para-medical activities, probably also indicates an ability to assign priorities and be self motivated.

A third candidate for canonical selection characteristics is communication ability. There is little doubt that successful medicine in most specialities requires not only technical knowledge and scientific expertise but also an ability to relate to patients and to empathize with them; and many medical accidents result not so much from technical failure but from failures of doctor–patient or inter-professional communication. If students are to be trained in communication skills (and there seems little doubt that they can be so trained and benefit from so doing (Simpson *et al.*, 1991, General Medical Council, 1991)), then it makes sense to concentrate training on those who already show a certain minimum level of ability. Whether medical school interviews can assess communications ability is not clear. Certainly those schools which do interview would claim it as an objective of interviews, and that seems more likely to be successful when 'ability to communicate' forms an explicit part of the ratings made for each interviewee.

Medical training

Just as one can define canonical characteristics for potential doctors, so one can define canonical skills and attitudes that should be learnt and encouraged during training. Clearly, a certain number of operations need to be seen, skills practised and patients clerked (although there is much variation in this (McManus *et al.*, 1993)). Can we discern any underlying themes of training that might make for safer medicine? Medical training is a vast area which we can hardly begin to consider here in its full richness. Defining clinical competence and the skills that underlie it is a complex and difficult task (Wakeford *et al.*, 1985; Neufeld and Norman, 1985), so that here we will consider just a few aspects of training which are often neglected, and which we believe may be especially relevant to errors and accidents.

Decision making. Doctors continually make decisions about patients, concerning diagnosis, aetiology, treatment, and management. It is surprising, therefore, how little effort has been put into *teaching* the process of decision making. Few medical schools have formal courses on diagnosis, or attempt to examine the process itself beyond mere exhortations to 'think', or the setting of example merely by apprenticeship. And likewise few doctors, at any level of specialization, are conversant with formal aids to decision making in medicine (Macartney, 1987; Thornton *et al.*, 1992; Schwartz and Griffin, 1986), despite the existence of computer software for aiding and making explicit the process. If faulty decision making is a key process in the genesis of medical accidents then the teaching of medical decision making should be an important part of the medical curriculum (Elstein, 1982; Elstein *et al.*, 1982).

The setting of objectives for training. Much emphasis in medical training is put upon the need for students to acquire a wide range of experience of different conditions. However, this experience is seldom actually assessed, with the result that by the end of their studies, medical students vary immensely in the experience they have gained of medical conditions and of practical procedures (Jolly and Macdonald, 1989; Dent *et al.*, 1990; McManus *et al.*, 1993). If experience is important then it should be possible to systematize the nature of the experience that particularly matters. The use of 'log-books' or 'checklists' (Hunskaar and Seim, 1984) can ensure that a sufficiently broad range and depth of experience is obtained, thereby setting minimum standards for all students. If combined with a curriculum that sets out specific learning objectives from such experiences, coupled with educational support in the form of tutorials and seminars, then experience will be far broader than that attained in the uncontrolled and unsystematic chance way that is typical of most medical schools.

The setting of objectives in examinations. There is a dictum in educational psychology that it is the mode of assessment which determines the method by which students study and learn. Conventional final medical examinations have many problems (Weatherall, 1991): they are often poorly organized, haphazard in their assessment of knowledge, ignore attitudes and skills, and typically do not observe the details of processes such as history-taking, examination, and diagnostic testing. The result is that students feel those processes are unimportant, and that the sole objective of learning is to be able to answer an examiner's questions, rather than to become self-critically competent. That effect is compounded by medical examinations being *summative* (i.e. occurring only at the end of the course for the purpose of assessment) rather than being *formative* (i.e. occurring repeatedly and with the intention of providing information to students on their progress), and they are typically *norm-referenced* (i.e. students are compared with their peers and a fixed percentage failed) rather than *criterion-referenced* (i.e. absolute standards are set, whereby in principle all students can pass or all can fail) (Turnbull, 1989). The development of OSCEs (Objective Structured Clinical Examinations), particularly if in the context of *Skills Labs* has helped students to concentrate on the structure of clinical tasks, ensuring that seemingly minor details, which are often concerned with safety, are not overlooked (Harden *et al.*, 1975; Black *et al.*, 1986; Newble *et al.*, 1981). A more widespread use of such assessment methods is likely to help increase safe behaviour by doctors.

The study of safety. It is rare for safety to be taught as an explicit part of clinical teaching. The result is that examinations rarely ask about safety *per se*, or about how to maximize it, how patients respond to accidents, and about the actions that should be taken by doctors when patients are injured. Now that medical schools are beginning to design core curricula, with specific educational objectives, it would seem reasonable to include such aspects within the objectives of medical courses. And then of course they would also require assessment.

Conclusions

In this chapter we have assessed the role that selection might play in reducing the numbers of medical accidents and mishaps. Although superficially attractive as an idea, our analysis of the potential reduction in accidents that might occur from selecting students principally for reduced accident-proneness is probably small, and that there would likely be other indirect effects on other criteria (such as academic ability) which could increase the rate of accidents. If selection is to have an influence upon accident rates then it must be an indirect one, through increasing the general competency of doctors. The potential effects of education and training upon accident rates are far greater than selection, and therefore the optimal form of selection should concentrate upon a small number of well-specified criteria which are likely directly to affect overall competency.

Three principal areas on which medical student selection might concentrate seem to be:

(1)   academic achievement as expressed in conventional public examination results;
(2)   study habits, and in particular a 'strategic' or 'deep' approach to learning;
(3)   ability to communicate, perhaps assessed in part by interviews, but probably capable of more formal, systematic assessment.

References

Alkov, R. A. and Borowsky, M. S. (1980). A questionnaire study of psychological background factors in US Navy aircraft accidents. *Aviation, Space and Environmental Medicine* **51**, 860–63.

Anon. (1992). Doctors to be: kingdom or exile? *Lancet* **340**, 1009.

Asher, J. J. (1972). The biographical item: can it be improved? *Personnel Psychology* **25**, 251–69.

Asher, J. J. and Sciarrino, J. A. (1974). Realistic work sample tests: a review. *Personnel Psychology* **27**, 519–33.

Bagg, D. G. (1970). A-levels and university performance. *Nature* **225**, 1105–1108.

Bale, R. M., Rickus, G. M., and Ambler, R. K. (1973). Prediction of advanced level aviation performance criteria from early training and selection variables. *Journal of Applied Psychology* **58**, 347–50.

Barmack, J. E. and Payne, D. E. (1961). Injury-producing private motor vehicle accidents among airmen: Psychological models of the accident-generating process. *Journal of Psychology* **52**, 3–24.

Becker, H. S., Geer, B., Hughes, E. C., and Strauss, A. C. (1961). *Boys in white: student culture in medical school*. University of Chicago Press, Chicago.

Biggs, J. B. (1978). Individual and group differences in study processes. *British Journal of Educational Psychology* **48**, 266–79.

Biggs, J. B. (1985). The role of meta-learning in study processes. *British Journal of Educational Psychology* **55**, 185–212.

Black, N. M. I., Urquhart, A. M., and Harden, R. M. (1986). The educational effectiveness of feedback in the objective structured clinical exam (OSCE). In *Newer developments in assessing clinical competence* (eds. I. R. Hart, R. M. Harden, and H. J. Walton) pp. 157–160. Heal Publications, Montreal.

Bloom, S. W. (1973). *Power and dissent in the medical school*. The Free Press, New York.

Choppin, B. (1979). Admission tests for admission to university: the British experience. In *The use of tests and interviews for admission to higher education* (ed. W. Mitter). National Foundation for Educational Research, Windsor.

Cobb, P. W. (1939). Automobile driver tests administered to 3363 persons in Connecticut, 1936–1937, and the relation of the test scores to the accidents sustained. Unpublished report to the Highway Research Board, Washington, DC.

Coles, C. R. (1985). Differences between conventional and problem-based curricula in their students' approaches to studying. *Medical Education* 19, 308–309.

Conger, J. J., Gaskill, H. S., Glad, D. G., Hassel, L., Rainey, R., and Sawrey, W. (1959). Psychological and psychophysiological factors in motor vehicle accidents. *Journal of the American Medical Association* 169, 1581–87.

Coombs, R. H. and Stein, L. P. (1971). Medical student society and culture. In *Psychosocial aspects of medical training* (eds. R. H. Coombs and C. E. Vincent). Thomas, Springfield, Illinois.

Craske, D. (1968). A study of the relationship between personality and accident history. *British Journal of Medical Psychology* 41, 399–404.

Dent, T. H. S., Gillard, J. H., Aarons, E. J., Crimlisk, H. L., and Smyth-Piggott, P. J. (1990). Pre-registration house officers in the four Thames regions: I. Survey of education and workload; II. Comparison of education and workload in teaching and non-teaching hospitals. *British Medical Journal* 300, 713–18.

Elstein, A. S. (1982). Decision making as educational subject matter. *Medical Decision Making* 2, 1–5.

Elstein, A. S., Rovner, D. R., and Rothert, M. L. (1982). A preclinical course in decision making. *Medical Decision Making* 2, 209–216.

Entwistle, N. and Waterston, S. (1988). Approaches to studying and levels of processing in university students. *British Journal of Educational Psychology* 58, 258–65.

Entwistle, N. J. and Wilson, J. D. (1977). *Degrees of excellence: the academic achievement game*. Hodder and Stoughton, London.

Entwistle, N., Hanley, M., and Hounsell, D. (1979). Identifying distinctive approaches to studying. *Higher Education* 8, 365–80.

Evans, L. (1991). *Traffic safety and the driver*. Van Nostrand Reinhold, New York.

Evans, G. W., Palsane, M. N. and Carrere, S. (1987). Type A behaviour and occupational stress: A cross cultural study of blue collar workers. *Journal of Personality and Social Psychology* 52, 1002–1007.

Fine, B. J. (1963). Introversion-extraversion and motor vehicle driver behaviour. *Perceptual and Motor Skills* 12, 95–100.

Fisher, R. A. (1973). *Statistical methods for research workers*, 14th edn. Hafner, New York.

Forbes, T. W. (1939). The normal automobile driver as a traffic problem. *Journal of General Psychology* 20, 471–474.

Fowler, B. (1981). The Aircraft Landing Test: an information processing approach to pilot selection. *Human Factors* 23, 129–37.

French, D. J., West, R. J., Elander, J., and Wilding, J. M. (1992). Decision-making style, driving style, and self-reported involvement in road traffic accidents. *Ergonomics*, in press.

General Medical Council (1991). *Interim document: Undergraduate medical education*. General Medical Council, London.

Ghiselli, E. E., Campbell, J. P., and Zedeck, S. (1981). *Measurement theory for the behavioral sciences*. Freeman, San Francisco.

Greenwood, M. and Yule, G. U. (1920). An enquiry into the nature of frequency distributions representative of multiple happenings, with particular reference to the occurrence of multiple attacks of disease or repeated accidents. *Journal of the Royal Statistical Society* 83, 255–79.

Greenwood, M., Woods, H. M., and Yule, G. U. (1919). *A report on the incidence of industrial accidents in individuals with special reference to multiple accidents, Report No. 4*. Industrial Fatigue Research Board, London.

Halpern, D. F. (1992). *Sex differences in cognitive abilities*, 2nd edn. Lawrence Erlbaum, Hillsdale, New Jersey.

Harano, R. M., Peck, R. C., and McBride, R. S. (1975). The prediction of accident liability through biographical data and psychometric tests. *Journal of Safety Research* 7, 16–52.

Harden, R. M., Stevenson, M., Downie, W., and Wilson, G. M. (1975). Assessment of clinical competence using objective structured clinical examination. *British Medical Journal* 1, 447–51.

Herriot, P. (1984). *Down from the ivory tower: graduates and their jobs*. Wiley, Chichester.

Hertz, D. (1970). Personality factor in automobile drivers. *Harefuah* 79, 165–67.

Hunskaar, S. and Seim, S. H. (1984). The effect of a checklist on medical students' exposures to practical skills. *Medical Education* 18, 439–42.

Irwin, J. O. (1964). The personal factor in accidents — a review article. *Journal of the Royal Statistical Society* 127, 438–51.

Jensen, R. S. (1982). Pilot judgement: training and evaluation. *Human Factors* 24, 61–73.

Jolly, B. C. and Macdonald, M. M. (1989). Education for practice: the role of practical experience in undergraduate and general clinical training. *Medical Education* 23, 189–95.

Jonah, B. A. (1986). Accident risk and risk-taking behaviour among young drivers. *Accident Analysis and Prevention* 18, 255–71.

Kaplan, H. B. and Pokorny, A. D. (1976). Self-derogation and suicide. II: Suicidal responses, self-derogation and accidents. *Social Science and Medicine* 10, 119–21.

Kravitz, R. L., Rolph, J. E., and McGuigan, K. (1991). Malpractice claims data as a quality improvement tool: I. Epidemiology of error in four specialities. *Journal of the American Medical Association* 266, 2087–92.

Landy, F. J. and Rastegary, H. (1989). Criteria for selection. In *Advances in selection and assessment* (eds. M. Smith and I. T. Robertson). Wiley, London.

Lipton, A., Huxham, G., and Hamilton, D. (1988). School results as predictors of medical school achievement. *Medical Education* 22, 381–88.

Loo, R. (1979). Role of primary personality factors in the perception of traffic signs and driver violations and accidents. *Accident Analysis and Prevention* 11, 125–27.

Macartney, F. J. (1987). Diagnostic logic. *British Medical Journal* 295, 1325–31.

Maccoby, E. E. and Jacklin, C. N. (1975). *The psychology of sex differences*. Oxford University Press, London.

McFarland, R. A. (1968). Psychological and behavioral aspects of automobile accidents. *Traffic Safety Research Review* 12, 71–80.

McGuire, F. L. (1972). A study of methodological and psycho-social variables in accident research. *JSAS Catalog of Selected Documents in Psychology* No. 195,

McKenna, F. P. (1983). Accident proneness: a conceptual analysis. *Accident Analysis and Prevention* 15, 65–71.

McManus, I. C. and Richards, P. (1984). An audit of admission to medical school: 1. Acceptances and rejects. *British Medical Journal* 289, 1201–4.

McManus, I. C. and Richards, P. (1986). Prospective survey of performance of medical students during preclinical years. *British Medical Journal* **293**, 124–27.

McManus, I. C. and Richards, P. (1989). Reliability of short-listing in medical student selection. *Medical Education* **298**, 147–51.

McManus, I. C., Maitlis, S. L., and Richards, P. (1989a). Shortlisting of applicants from UCCA forms: the structure of pre-selection judgements. *Medical Education* **23**, 136–46.

McManus, I. C., Richards, P., and Maitlis, S. L. (1989b). Prospective study of the disadvantage of people from ethnic minority groups applying to medical schools in the United Kingdom. *British Medical Journal* **298**, 723–26.

McManus, I. C., Tunnicliffe, N., and Fleming, P. R. (1990). The independent effects of intelligence and educational achievements in predicting final examination success (Abstract). *Medical Education* **24**, 181–84.

McManus, I. C., Richards, P., Winder, B. C., Sproston, K. A., and Vincent, C. A. (1993). The changing clinical experience of British medical students. *Lancet*, **341**, 941–4.

Mawhinney, B. S. (1976). The value of ordinary and advanced level British school-leaving examination results in predicting medical students' academic performance. *Medical Education* **10**, 87–89.

Maycock, G., Lockwood, C. R., and Lester, J. F. (1991). *The accident liability of car drivers. TRRL Research Report 315.* Traffic and Road Research Laboratory, Crowthorne, Berkshire.

Merton, R. K., Reader, G. G., and Kendall, P. L. (1957). *The student physician.* Harvard University Press, Cambridge, Massachusetts.

Neufeld, V. R. and Norman, G. R. (1985). *Assessing clinical competence.* Springer, New York.

Newble, D. I. and Gordon, M. I. (1985). The learning style of medical students. *Medical Education* **19**, 3–8.

Newble, D. J., Hoare, J., and Elmslie, R. G. (1981). The validity and reliability of a new examination of the clinical competence of medical students. *Medical Education* **15**, 46–52.

Newbold, E. M. (1926). *A contribution to the study of the human factor in the cause of accidents, Report No. 34.* Industrial Health Research Board, London.

Owens, W. A. (1976). Background data. In *Handbook of industrial and organisational psychology* (ed. M. D. Dunnette). Rand McNally, Chicago, Illinois.

Owens, W. A., Glennon, J. R., and Allbright, L. E. (1966). *A catalog of life history items.* Richardson Foundation, Greensnoro, North Carolina.

Perry, A. R. (1986). Type A behaviour pattern and motor vehicle driver behaviour. *Perceptual and Motor Skills* **63**, 875–78.

Pestonjee, D. M. and Singh, U. B. (1980). Neuroticism-extraversion as correlates of accident occurrences. *Accident Analysis and Prevention* **12**, 201–4.

Reilly, R. R. and Chao, G. T. (1982). Validity and fairness of some alternative employee selection procedures. *Personnel Psychology* **35**, 1–62.

Richards, P., McManus, I. C. and Maitlis, S. (1988). Reliability of interviewing in medical student selection. *British Medical Journal* **296**, 1520–21.

Richardson, I. M. (1980). Examination performance and the future careers of Aberdeen medical graduates. *Medical Education* **14**, 356–59.

Robertson, I. T. and Smith, M. (1989). Personnel selection methods. In *Advances in selection and assessment* (eds. M. Smith and I. T. Robertson). Wiley, London.

Rolph, J. E., Kravitz, R. L., and McGuigan, K. (1991). Malpractice claims data as a quality improvement tool: II. Is targeting effective? *Journal of the American Medical Association* **266**, 2093–97.

Rose, G. and Barker, D. J. P. (1986). *Epidemiology for the uninitiated*, 2nd edn. British Medical Journal, London.

Savage, R. D. (1972). An exploratory study of individual characteristics associated with attainment in medical school. *British Journal of Medical Education* **6**, 68–77.

Schmidt, F. L., Ones, D. S., and Hunter, J. E. (1992). Personnel selection. *Annual Review of Psychology* **43**, 627–70.

Schuman, S. H., Pelz, D. C., Westervelt, F. H., and Quinn, J. (1967). Young male drivers: Impulse expression, accidents and violations. *Journal of the American Medical Association* **200**, 1026–30.

Schwartz, S. and Griffin, T. (1986). *Medical thinking: The psychology of medical judgement and decision making.* Springer, New York.

Selzer, M. L. and Vinokur, A. (1974). Life events, subjective stress and traffic accidents. *American Journal of Psychiatry* **13**, 903–6.

Shaw, L. and Sichel, H. S. (1971). *Accident proneness.* Pergamon Press, Oxford.

Simpson, M., Buckman, R., Stewart, M., Maguire, P., Lipkin, M., Novack, D., and Till, J. (1991). Doctor-patient communication: the Toronto consensus statement. *British Medical Journal* **303**, 1385–87.

Singh, A. P. (1978). Neuroticism, extraversion and accidents. *India Psychological Review* **16**, 41–45.

Sloan, F. A., Mergenhagen, P. M., Burfield, W. B., Bovbjerg, R. R., and Hassan, M. (1989). Medical malpractice experience of physicians. *Journal of the American Medical Association* **262**, 3291–97.

Sloan, F. A., Mergenhagen, P. M., Burfield, W. B., Bovjberg, R. R., and Hassan, M. (1991). Medical malpractice experience of physicians: predictable or haphazard? *Journal of the American Medical Association* **262**, 3291–97.

Stuart, J. C. and Brown, B. M. (1981). The relationship of stress and coping ability to incidence of diseases and accidents. *Journal of Psychosomatic Research* **25**, 255–60.

Suchman, A. E. (1970). Accidents and social deviance. *Journal of Health and Social Behaviour* **11**, 14–15.

Thornton, J. G., Lilford, R. J. and Johnson, N. (1992). Decision analysis in medicine. *British Medical Journal* **304**, 1099–1102.

Tomlinson, R. W. S., Clack, G. B., Pettingale, K. W., Anderson, J., and Ryan, K. C. (1977). The relative role of 'A' level chemistry, physics and biology in the medical course. *Medical Education* **11**, 103–08.

Tooth, D., Tonge, K., and McManus, I. C. (1989). Anxiety and study methods in pre-clinical students: causal relation to examination performance. *Medical Education* **23**, 416–421.

Turnbull, J. M. (1989). What is ... Normative versus criterion-referenced assessment? *Medical Teacher* **11**, 145–50.

Wakeford, R., Bashook, P., Jolly, B., and Rothman, A. (1985). *Directions in clinical assessment: Report of the first Cambridge Conference.* Cambridge University School of Clinical Medicine, Cambridge.

Weatherall, D. J. (1991). Examining undergraduate examinations. *Lancet* **338**, 37–39.

West, R. J., Elander, J., and French, D. J. (1992a). Mild social deviance, type-A personality and decision making style as predictors of self-reported driving style and traffic accident risk. (Unpublished report).

West, R. J., Elander, J., and French, D. J. (1992b). Decision making, personality and driving style as correlates of individual accident risk. Report presented to Road Safety Division, Transport and Road Research Laboratory, 1992. (Unpublished report).

Whitlock, F. A., Stoll, J. R., and Rekhdahl, R. J. (1977). Crisis, life events and accidents. *Australian and New Zealand Journal of Psychiatry* **11**, 127–32.

Wilson, R. and Greensmith, J. (1983). Multivariate analysis of the relationship between drivometer variables and drivers' accident, sex and exposure status. *Human Factors* **25**, 303–12.

7

The training of junior doctors

MAEVE ENNIS

Juniors are the doctors most frequently implicated in medical accidents. This is not surprising given the ratio of junior to senior doctors in most hospital departments. However, studies have also suggested that it is not numbers alone that account for the greater involvement of juniors in these incidents, but that it is also the inadequacy of Senior House Officer training (Buck *et al.*, 1990; Ennis and Vincent, 1990; Confidential Enquiry into Maternal Deaths, 1991). If this is true then better training could eliminate many of these accidents.

The General Medical Council and many of the Royal Colleges have laid down recommendations for the training of specialists in the UK. Medical education in Britain is divided into a two-year preclinical period of academic work devoted to studying basic medical science. Students then enter a three-year clinical period in which they are attached to a series of different specialties. Some formal academic study continues but students are mostly taught by clinicians in the course of their routine work. This is followed by a one-year period of general clinical training, known as the preregistration year. Completion of this period and a professional or qualifying exam such as MBBS leads to full registration. During this period the trainee doctor will, under supervision and without any real responsibility for patients, become familiar with clinical medicine and patient care.

The two to three years following full registration is occupied by basic specialist training, during which period a doctor acquires increased but supervised responsibility for patient care and is expected to develop a wide range of skills needed for practice in the speciality he or she has chosen. Equivalent experience is required for those specializing in general practice. Higher specialist training of a period of three to five years follows, at the end of which a doctor is expected to have completed training and be ready for consultant posts. This chapter will concern itself primarily with the two- to three-year postregistration training period during which the doctor works as a Senior House Officer (SHO).

Postgraduate medical education in Britain has been criticized by SHOs themselves among others (Dent *et al.*, 1990; Ennis 1991). The training of SHOs has been described by one author as a 'brutal enterprise' in which the junior doctors were likened to apprentices to a trade rather than students of a profession (Roberts, 1991). That author points to the long hours worked, the uncertain futures, the dependence on senior staff rather than merit for advancement, and the fluctuations of current government policies in a changing National Health Service as evidence for this. There are other authors who agree that training in the NHS has developed on an apprenticeship model with service providing the framework for training. A report from the Council of Postgraduate Medical Education believes that service content and manpower structure have evolved in response to patient need and professional demand with little concern for their training potential (CPME, 1987). However, there are also those who defend this system, arguing that training in this 'hands on' fashion results in doctors who are knowledgable, skilled, and dedicated. However, publication of the CEPOD report (1987) and the Confidential Enquiry into Maternal Deaths (1991) have called attention to the relationship between outcome and the level of experience of those attending patients. They cite inexperience and lack of adequate supervision of junior staff as factors in the incidents reported.

The President of the Royal College of Physicians, in a foreword to a report of a standing committee of the Royal College, suggests that both views may be pertinent. He notes that the complexities of modern clinical practice due to the advance of modern medical knowledge call for a balance between the acquisition of broad-based general medical knowledge and experience and a high degree of expertise and sophistication such as that derived from the specialist training advocated by the Royal Colleges (RCP Standing Committee of Members, 1986). Several recent investigations into the training of SHOs would suggest that this balance is not being achieved (Grant, 1989; Reeve and Bowman, 1989; Crawley and Levin, 1990; Morris *et al.*, 1990; Ennis, 1992; Kelly and Murray, 1991; Blunt, 1991). The majority of these studies identified similar problems such as trial-and-error learning, lack of feedback, lack of formal teaching, and unsupervised clinical practice, and all do indeed point to an apprenticeship model of training.

Formal training

Our own studies in the training of obstetric SHOs found that their perception was that they were inadequately trained for the work they were expected to carry out. The study looked at seven hospitals: four teaching hospitals and three District General hospitals. It focused particularly on specific problem areas that were the areas of most concern raised by a study of cases that had come to litigation (Ennis and Vincent, 1990). These areas were: training in the use of forceps, interpretation of cardotocographic traces (CTGs), and the lack of supervision from more senior staff. It was found that at the end of six months' training in obstetrics, 24 per cent of

SHOs stated that they had received no training in the use of forceps. Furthermore, of those who had been trained, 40 per cent felt that their training was inadequate. A similar result was found for CTGs, with 50 per cent of SHOs reporting no training in their interpretation. They said they depended on other doctors and midwives to detect their errors. Not surprisingly, most of the SHOs interviewed felt that this learning on a trial-and-error basis was inadequate. With one to two hours of formal teaching a week, and less in some hospitals, their responses are easily understood. The doctors in this study were also questioned about the level of supervision they considered they had received. Few reported that their work had been directly supervised by a consultant; most supervision being carried out by registrars. Furthermore, 20 per cent said they had had received little or no supervision during their six months training. In summary, the majority of these trainee specialists felt they were adequately prepared for the day-to-day work they were expected to carry out on labour wards and delivery suites (Ennis, 1991).

Similar studies have been carried out in other areas. In Accident and Emergency, for example, a survey of SHOs at eight different hospitals who had at least three months experience revealed that while all sixty doctors in the study had received one to two hours teaching each week from senior staff, eighteen had not received any teaching on multiple or minor injuries, nine had received none on resuscitation and thirty-seven had received none on spinal injuries (Morris *et al.*, 1990). The authors of this study concluded that their results highlighted important deficiencies in the training of junior accident and emergency staff which needed urgent attention. They suggest that a comprehensive introductory course should be provided, with formal and informal teaching concentrating on the most important and life-threatening conditions (Morris *et al.*, 1990). Yates and Wakeford (1985), in an earlier study, had shown that initial clinical and practical knowledge did not match the demands of the post and that three months after appointment some doctors considered themselves unready to assume responsibility for important and frequently presenting clinical conditions. These authors pointed out that the doctors in their study were those who every day were making decisions unaided as to whether or not to discharge patients, assess priorities, deal with crises, and treat a variety of minor medical and surgical conditions. The majority of juniors in the study expressed dissatisfaction with the teaching they received. The authors concluded that previous experience does not adequately prepare SHOs for the day-to-day problems they encounter in A & E departments or for the practical procedures they are commonly expected to carry out (Yates and Wakeford, 1985). While the authors do not comment on the amount of teaching or training the SHOs in their study were receiving while in post, they do suggest that as SHOs in A & E are expected to assume a great deal of responsibility for patient care from the onset, they need an intensive introductory training course at the beginning of their attachment and before taking on the responsibility.

Studies in other specialties have led to similar conclusions. A study of hospital training for General Practice revealed that there would seem to be serious educational and training deficiencies in hospital posts which are used in vocational training schemes of GPs. The authors suggest that these deficiencies also exist for other specialist trainees (Reeve and Bowman, 1989). They also found that trainees had little or no formal teaching: in 37 per cent of posts no formal teaching at all was received and where formal teaching was carried out it occupied no more than one to two hours a week.

Informal training

A study conducted by the Office of Manpower Economics support the conclusions of the studies already cited. It was found that overall formal educational activities occupied only 1.68 per cent hours of an SHO's average working week (Review Body on Doctors and Dentists Remuneration, 1987). This does not mean that training is not occurring outside these 1.68 per cent hours or indeed the one to two hours a week reported in the above studies: informal teaching by example and discussion may well be going on. However, when trainees on a GP vocational training scheme were questioned about informal teaching, 28 per cent said they had received none whatsoever and in the same study 37 per cent of trainees said they had received no formal training (Reeve and Bowman, 1989). If we are to take these results at face value then a sizeable proportion of trainees from this study will have received no instruction at all on diagnosis and patient care. But, as the authors point out, informal teaching is by its nature often a covert activity unrecognized by trainees for what it is, and they feel some caution should be expressed when interpreting these results.

The value of informal teaching has been emphasized elsewhere (CPME, 1987; GMC, 1989) but to my knowledge no empirical work on its frequency or validity as a method of training has been carried out. Anecdotal evidence from consultants suggest that they hold differing views from SHOs on its efficiency. Ward rounds and mortality and morbidity meetings are seen by consultants as valuable learning tools, where experienced clinicians discuss and debate the practice of day-to-day hospital medicine at all levels of severity in an atmosphere where juniors are free to comment and question.

A communication study carried out in obstetrics (Ennis, 1992) suggests that, in this speciality at least, this assessment of informal teaching by consultants is not shared by SHOs. In this study, thirty-nine SHOs were questioned about their communication with senior staff. Eighty per cent said they had little contact with consultants and two reported that over a six-month period their consultant had never spoken to them. Forty per cent stated that they would not voice disagreement with a more senior doctor either about patient care or a prescribed treatment. Many said they were discouraged from asking questions — sometimes by being ridiculed when they did so. Female SHOs reported that this was sometimes accompanied by sexist comment. My own, albeit limited, experience as an observer at morbidity and mortality meetings, in more than one specialty (I have attended some 20 meetings), suggests that while many are conducted in an atmosphere that is very conducive to learning, some are also used to scapegoat juniors when things go wrong. At some of the meetings I attended, senior staff responsible for training were conspicuous by their absence.

Study leave

Study leave is also seen as an important aspect of training (CPME, 1987; GMC 1989) but in a study carried out on vocational training for GPs in one region (Reeve and Bowman, 1989) it was applied for by only 37 per cent. The authors state that few trainees were encouraged to see it as an opportunity to broaden their experience and in some posts they were actively discouraged from taking it. A large study which looked at all specialties in another region found that of those applying for study leave, 63–75 per cent were granted it. However, they do not report what proportion of the SHOs surveyed actually applied (Grant and Marsden, 1988). Most study leave was taken to attend a course for postgraduate examinations. Many SHOs were made to feel that study leave was a favour rather than a right. As study leave is not timetabled into SHO posts as an essential part of their duties, many felt guilty about applying because their SHO colleagues had to cover for them during their absence. While the SHOs themselves felt that they were not, on the whole, encouraged to take study leave and that consultants held generally negative views about it, most consultants reported that while they find study leave inconvenient they encourage their SHOs to apply for it. However, consultants appeared to be unaware that their juniors often took annual leave rather than study leave to prepare for examinations.

Service or training

The above brief review of the literature paints a depressing picture of the training of SHOs and suggests that the line between service and training for SHOs is ill-defined and it supports the perception of current practice, as an apprenticeship 'hands-on model' of training with insufficient planned teaching, either formal or informal. SHOs consider that they provide more service than they receive training, a state of affairs they find unsatisfactory. In the study by Grant and Marsden (1988), registrars and senior registrars agreed with SHOs on this point and their views were shared by some consultants. However, many consultants did not believe this was a correct view. Other authors report the dissatisfaction of SHOs with the high service content of their job which leaves little room for learning more about practical procedures (Walker, 1991). Marteau *et al.* (1990) have demonstrated that experience is not a substitute for planned training, and that juniors need formal training in skills and feedback on their performance for this to improve. Our own study on the mismatch between skill and confidence in reading fetal heart traces supports this conclusion (Ennis *et al.*, 1992). Garrud (1990) further reported that SHOs welcomed feedback on their work performance. In a study of 106 SHOs who had graduated from one medical school and who were working in various specialties he found that about three quarters had no formal appraisal or feedback on their performance, some only finding out that consultants held negative views of their work when they asked for references to support the applications for their next post. All of those who received feedback found it constructive and useful and some said they would have liked more. They looked for positive as well as negative feedback; and for appraisal of their performance to include an opportunity to discuss clinical decisions and policy as well as feedback on their skill. SHOs have been shown to maintain inappropriate self-confidence (Haas and Shaffir, 1982) and without appraisal they may adapt a personal strategy of apparent confidence to conceal their own uncertainties and inadequacies. Appropriate feedback and training would assist in associating confidence more appropriately with actual levels of performance.

Achieving better-trained SHOs

The studies reviewed above have shown that there is a need for improved training standards for SHOs. It is difficult in the face of the evidence to sustain the traditional belief that junior doctors can learn effectively from 'hands on' experience. There is no doubt that this is becoming more widely accepted, and that the development of new approaches to training requires more attention.

HOURS OF WORK AND WORKLOAD

It is impossible to look at the training needs of junior hospital doctors without looking at hours of work and workload. It is generally agreed that the hours worked, on average 90 hours a week, are too long and the detrimental effects of this are dealt with elsewhere in the book (see Chapter 9). Steps are being taken to reduce this (Department of Health, 1990). However, it cannot be achieved without creating more posts at a higher level and although the Government has expressed a commitment to this, the resultant reality may fall far short of any ideal expressed by doctors, whether senior or junior. The Department of Health proposes to reduce the long hours worked by junior doctors by creating new consultant posts in 'hard pressed' specialties such as Obstetrics and Gynaecology. The aim is to reduce not just the hours but also the workload on juniors and to free more of consultant's time for the training of juniors. When the fine detail of this proposal was examined we found that with 280 district general and teaching hospitals and 1600 consultants in England and Wales alone, the proposed 2 per cent increase in consultant posts would not make much difference, and it is not at all clear that this initiative will lead to any effective changes in working practices.

Protocols

It is believed by most researchers in the area that one of the prime needs of doctors in training is the need for guidelines or protocols. These are standardized specifications for care, either for using a procedure or for managing a particular clinical problem. In many hospital departments in the UK, very good protocols are in use but many others they are not. The arguments raised against written protocols or guidelines are that failing to conform to them in a special circumstance would lay doctors open to litigation and, further, that they remove clinical freedom from the practitioner. However, studies carried out in the USA have shown that widespread adherence to guidelines have induced some improvements in medical care and in some areas have decreased the number of claims against doctors. For example, the anaesthesiologists at Harvard University instituted what they call practice guidelines several years ago. Many of the these concern simple precautions such as the requirement that anaesthesia personnel be continuously present in the operating theatre. They have found that over the last three years the number of claims based on anaesthesiology errors have decreased substantially and so have injuries to patients (Brennan, 1991). In a study in the North of England of 236 cases for which damages had been paid, it was found that 33 were directly attributable to failure to follow recognized safety procedures (Woodyard, 1990). It would therefore appear that protocols and guidelines can lead to both better quality care and less litigation, and it has been particularly emphasized that they are effective in preventing errors of omission (Lilford *et al.*, 1992).

Introductory courses

Training for a particular specialty has to start when SHOs take up their first post; it cannot be assumed that SHOs on rotation will come into a unit with skills relevant to that specialty. Several authors have suggested that an introductory course at the beginning is the most efficient method of achieving competence. This should comprise formal and informal teaching and it could be a period when SHOs can familiarize themselves with unit procedures without feeling they are expected to assume responsibility (Yates and Wakeford, 1985; Morris *et al.*, 1990; Pogmore, 1992).

Pogmore (1992), who is chairman of the Hospital Recognition Committee for Obstetrics and Gynaecology, has suggested that midwives have a role to play in the training of SHOs. Experienced midwives have skills that obstetric SHOs need to acquire. He suggests that midwives should be encouraged to involve SHOs in the management of women in labour and that the SHO should accept the combined role of colleague and trainee. He further suggests that protocols for the management of clinical problems in the labour ward should be agreed between midwives and consultants and that consultants come to the labour ward regularly and use clinical events to teach the SHOs and check that the protocols are being followed.

The NHS is in a state of flux. The government White Paper, *Working for patients*, is committed to improved quality of care. However, there is some disquiet among junior doctors that it will negate educational reform. They believe that the 'internal marketplace' of hospital trusts will create more incentive to use junior doctors as cheap labour (Roberts, 1991). This disquiet is shared by the Standing Committee on Postgraduate Medical Education. This committee considers that the content of the White Paper and the way it has been launched will threaten eduction and training; and that many of the proposals contained in the document are time-consuming and will deflect consultants from their role as trainers.

Current changes taking place in the NHS are profound and a new, more open culture is emerging. Many of the Royal Colleges and the GMC are discussing restructured proposals for training. However, managers in the health service also have a role to play. The introduction of crown indemnity for medical and dental negligence or malpractice has meant that senior management are more involved in negligence cases. Districts need to set up enquiries when things go wrong. When cases occur because inexperienced junior doctors are working without adequate help or training, their managers must look at staffing levels of senior doctors and patient throughput. There is an increasing trend in districts to 'buy in' risk management and it is hoped that managers will see the correlation between good risk management and training and to see that they too are accountable for ensuring that SHOs are properly trained.

Finally, there is a need for more research into link between the training of junior doctors and medical accidents; there is a paucity of studies in this area. This should be funded by the Department of Health.

References

Blunt, S. M. (1991). Training in obstetrics. *British Medical Journal*, **303**, 1416.

Brennan, T. A. (1991). Practice guidelines and malpractice litigation: collision and cohesion. *Journal of Health Politics, Policy and Law* **16**, 65–87.

Buck, N., Devlin, H. B., and Lunn, J. N. (1990). *Confidential enquiry into perioperative deaths*. Nuffield Provincial Hospitals Trust, London.

Confidential Enquiry into Maternal Deaths (1991). *Report on confidential enquiries into maternal deaths in the United Kingdom 1985–1987*. HMSO, London.

CPME (1987). *The problems of the senior house officers*. Council for Postgraduate Medical Education in England and Wales, London.

Crawley, H. S. and Levin, J. B. (1990). Training for general practice: a national survey. *British Medical Journal* **300**, 911–15.

Dent, T. H. S., Gillard, J. H., Aarons, E. J., Crimlink, H. L., and Smyth-Pgott, P. J. (1990). Preregistration House Officers in the four Thames regions: 1. Survey of education and workload. *British Medical Journal* **300**, 713–15.

Department of Health, Welsh Office. DHSS Northern Ireland. Scottish Home and Health Department (1990). *Heads of agreement: ministerial group on junior doctors' hours*. DoH, London.

Ennis, M. (1991). Training and supervision of obstetric senior house officers. *British Medical Journal* **303**, 1442–43.

Ennis, M. (1992). *Communication on the labour ward* (unpublished manuscript).

Ennis, M. and Vincent, C. A. (1990). Obstetric accidents: a review of 64 cases. *British Medical Journal* **300**, 1365–67.

Ennis, M., Reid, P., McClelland, A., Fraser, R., and Audley, R. J. (1992). *Mismatch between skill and confidence in reading fetal heart traces* (unpublished manuscript).

Garrud, P. (1990). Counselling needs and experience of junior hospital doctors. *British Medical Journal*, **300**, 445–47.

GMC. (1987). *Recommendations on the training of specialists*. General Medical Council: Education Committee, London.

Grant, J. and Marsden, P. (1988). *Senior house officer training in South-East Thames*. Report for: South-East Thames Health Regional Authority.

Haas, J. and Shaffir, W. (1982). Ritual evaluation of competence. The hidden curriculum of professionalization in an innovative medical school program. *Work and Occupations* **9**, 131–54.

Kelly, D. and Murray, T. S. (1991). Twenty years of vocational training in the west of Scotland. *British Medical Journal*, **302**, 28–30.

Lilford, R. J., Kelly, M., Baines, A., Cameron, S., Cave, M., Guthrie, K., and Thornton, J. (1992). Effect of using protocols on medical care: randomised trial of three methods of taking an antenatal history. *British Medical Journal* **305**, 1181–86.

Marteau, T. M., Wynne, G., Koyce, W., and Evans, T. R. (1990). Resuscitation: experience without feedback increases confidence but not skill. *British Medical Journal* **300**, 849–50.

Morris, F., Cope, A., and Hawes, S. (1990). Training in accident and emergency: views of senior house officers. *British Medical Journal* **300**, 165–66.

Pogmore, J. R. (1992). Role of senior house officers in the labour ward. *British Journal of Obstetrics and Gynaecology* **99**, 180–81.

Reeve, H. and Bowman, A. (1989). Hospital training for general practice: views of trainees in the North Western region. *British Medical Journal* **298**, 1432–36.

Review Body of Doctors and Dentists Remuneration. (1987). *Seventeenth Report*. HMSO, London.

Roberts, J. (1991). Junior doctors' years: training, not education. *British Medical Journal* **302**, 225–28.

RCP Standing Committee of Members (1986). *Training to be a physician*. Royal College of Physicians, London.

Walker, A. (1991). Teaching junior doctors practical procedures. *British Medical Journal* **302**, 306.

Woodyard, J. (1990). Facing up to errors. *The Health Service Journal* **100** (5194), 468–69.

Yates, D. W. and Wakeford, R. (1985). The training of junior doctors for accident and emergency work: a case for urgent treatment? *Injury* **14**, 456–60.

8

Would decision analysis eliminate medical accidents?

JACK DOWIE

Introduction

In medical terms, to apply the term 'accident' is to make a diagnosis. And, as is often the case in medicine, the making of this particular diagnosis has a major impact on the 'treatment' that follows. The power to make the diagnosis 'accident' is therefore highly prized because of its implications and repercussions for the various parties to the event concerned.

What are the characteristics of this diagnosis? What 'tests' are applied to determine its appropriateness? In both ordinary language and its legal formalization to call something 'an accident' is to imply that it was largely, if not completely, unintentional, unforeseeable — and harmful. If the outcome of an action or event were not harmful it is unlikely to be designated an accident, whether or not it was unintentional or unforeseeable. And if it *was* harmful, but either intentional or foreseeable, then it is also unlikely to be designated an accident — except as a strategic move by a party whose interests lie in such a labelling.

This chapter will be mainly concerned with exploring whether, in the context of medical practice, these three 'tests' of whether an event was an accident could survive the introduction of formal decision aids and methods, in particular the specific method called clinical decision analysis. And, if they could not, what would follow for the notion of 'medical accidents'.[1]

The cognitive basis of medical action

The medical encounter is, or should be, the setting *par excellence* for conscious and rational choice. Views differ as to its proper characterization, both descriptively and normatively. In one extreme view, the patient hands over complete power to the doctor as agent, on the understanding that the

---

[1] The sections outlining cognitive continuum theory and decision analysis are similar to those appearing in my other forthcoming papers in this area, notably 1993a and 1993b. Readers are invited to consult these for elaboration of some of the arguments contained in this chapter.

doctor will take the decision and follow the course of action that is in the patient's best interest. The patient's role is to supply such information as the doctor requests and to accept such information as the doctor thinks fit to offer them. In the other extreme view, it is the patient who is the decision maker. He or she is engaged in the encounter to acquire the information possessed by the doctor (as technical consultant). Having made the decision the patient often also needs the doctor (as licensed practitioner) in order to implement it, for example by providing a drug prescription or performing an operation.

Fortunately we do not need to resolve the debate as to who does what, or who should do what, in the medical encounter, since it is the avowed intention of all parties to do what is in the patient's best interest after 'serious thought'.

Intending, foreseeing, and evaluating (i.e. assessing degree of harm or benefit) are all fundamentally cognitive operations, and genuine decision aids and methods are directed at the cognitive basis of medical practice. In brief, they affect the way information is processed by the practitioner (and the other parties involved in a decision) and not simply the way information is provided and/or presented to them for use. So an 'aid' that provides a practitioner with more or better data, or presents data in a more legible or easily interpretable way, is a 'data or information aid', not a decision aid. Such an aid may, and hopefully will, improve the inputs into the decision. And it will perhaps improve the decision as a result of improving the inputs, in the same way that better ingredients may improve the cake. But it still leaves control of the process of decision making in the hands (mind) of the people. A genuine decision aid or method influences the way the decision problem is framed and how it is tackled, including in many cases defining what is relevant data and how it is to be processed.

The use of a genuine decision aid or method, and especially the adoption of a decision analytic approach, has one very obvious but very easily overlooked effect. It is made inescapable that a decision is being made. It is not uncommon, when raising the question of their judgement and decision making with medical practitioners, to find them unfamiliar with, and somewhat resistant to, the idea that making judgements and decisions is a separate, or separable, part of their task. Many simply see themselves involved in an undifferentiated flow of activity called 'practising medicine'. The use of decision aids establishes the existence of decision making as an activity distinct from other aspects of practice, especially the performance of the actions (or inaction) following from the decision. Having been made distinct, medical judgements and decisions accordingly become open to assessment. The nature and quality of 'clinical judgement' becomes a subject for investigation and evaluation. The term can no longer be used as professional code to signify the closure of debate.

It is largely the fact that decision aids require conscious and continuing acknowledgement that decisions are the basis of medical actions which means their dissemination will transform the way in which all outcomes of medical practice, including 'medical accidents', are viewed. In the author's view, the concept of 'accident' can retain little attraction once the decisional basis of medical practice has been fully established and exposed to all the parties to medical encounters. If decision analysis were to be accepted as the normative process for medical decision making it would indeed be the end of 'medical accident' as a diagnostic label, though not of course of the sorts of undesired outcomes that typically provoke its use. Like other diagnostic labels — especially the explanation-free 'syndromes' — its demise would be delayed only by individuals who found it a source of emotional comfort or, as already hinted, by groups who felt they benefited from its availability.

Despite its lack of penetration of real-world medicine (Thornton *et al.*, 1992; Böckenholt and Weber, 1992), decision analysis is the normatively best-grounded method of aiding and making decisions and poses the greatest challenge to current thinking on 'medical accidents'. It therefore seems optimal, in this short chapter, to concentrate on its implications. Fuller elaborations of other judgement and decision aids and methods, such as 'bootstrapping', 'Bayesian scoring systems' and 'knowledge-based expert systems', all of which have similar, albeit narrower or weaker, implications for 'accidents', can be found elsewhere (Dowie, 1992).

Before outlining what is involved in decision analysis and establishing why it has such major implications, it will be useful to set the discussion in the wider framework for thinking about judgement and decision making that is provided by cognitive continuum theory.

Modes of judgement and decision making

Kenneth Hammond's cognitive continuum theory (Hamm, 1988) proposes a two-dimensional framework for studying judgement and decision making (Fig. 8.1). The cognitive dimension (from which the theory gets its name) focuses on how we think about the task and runs from 'highly intuitively' to 'highly analytically'. The task dimension focuses on how the task presents itself and runs from 'very ill-structured' to 'very well-structured'. The characteristics which determine how ill or well a task is structured are numerous and it is sufficient for our purposes to illustrate them. A GP's domiciliary visit to a previously unseen patient will confront him or her with a fairly ill-structured task in that he or she will have many cues of a largely non-quantitative nature becoming available almost simultaneously. On the other hand, a consultant examining the latest test results for an inpatient who has been thoroughly worked up over many days, is presented with a relatively much more structured task in that he or she will have few cues of a largely quantitative nature becoming available sequentially.

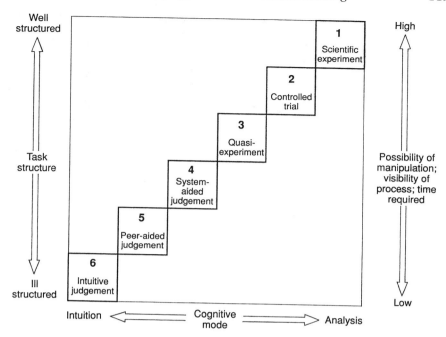

Figure 8.1 The cognitive continuum.

While the two dimensions are continuous, Hammond's framework suggests that six broad modes of inquiry can usefully be identified when we locate practices on both dimensions.

At one extreme we find the classic scientific experiment of chemistry and physics where one adopts a highly analytic approach to a very well structured task. Progressively less well structured tasks induce the progressively less analytical modes of the controlled trial and the quasi-experiment (the latter being particularly familiar to epidemiologists). All three of these modes provide much of the basic knowledge used by clinical practitioners, but they are not ones which clinicians practice as clinicians. The remaining three modes are the most clinically relevant ones.

Furthest removed from the scientific experiment is the mode labelled intuitive judgement and decision making, which involves using little analysis in approaching an ill-structured task. It needs to be emphasized that the term 'intuitive' is used non-pejoratively and that lack of analysis is not meant to imply that experience and knowledge are not being drawn on in some way. 'Pattern recognition' is the prototypical example of mode 6 practice. A more analytical approach to a somewhat better structured task takes us to the peer-aided judgement and decision making typical of the hospital setting and some group practices in primary care. The need to articulate the case to colleagues is a prime reason for the higher level of analysis characteristic of this mode.

The remaining mode (mode 4 in the figure) is the prime focus here, in that it is the one in which 'clinical decision analysis' and most other decision aids are located. It is called system-aided judgement and decision making.

One particular aid which comes into use for the first time in mode 4 judgement and decision making, and is a key feature of decision analysis, is the number system. Much medical discourse gives the impression that numbers are to be used when they are not. 'We must take all the circumstances into account' and 'give all the relevant considerations proper weight' and 'strike the right balance' and 'get things in proportion'.

Despite this, it is not numbers but words or phrases — charitably called 'verbal quantifiers' — that are the common form in which assessments of chance or frequency are communicated among clinicians. 'Rare', 'probable', 'rather likely', 'fairly unlikely', and 'a good chance' are typical. The same applies where assessments of outcome severity or desirability are concerned. We constantly encounter 'severe', 'moderate', 'mild', 'reasonably good', 'excellent', and so on. Many studies suggest that the degree of interpersonal agreement on the meaning of these terms is low.

At mode 4 the metaphorical use of the concepts of 'weighing', 'balancing', 'taking things into account', and 'establishing a sense of proportion' is replaced with actual, numerical, weighing, balancing, and — literally — taking into a count and establishment of proportions. And attempts to 'verbally quantify' orders of magnitude, either of frequency or severity, are replaced by attempts to numerically quantify them.

Having explicitly and numerically quantified what is otherwise implicitly and covertly quantified, the mode 4 practitioner is in a position to bring to bear (in the patient's best interest) another set of valuable 'aids' that have been the focus of sustained development for some centuries. These are the normative principles of mathematical and statistical reasoning. In a word, the principles of calculation, which, it unfortunately needs to be emphasized, are just as useful in identifying and pursuing 'good' ends (e.g. a patient's well-being) as in identifying and pursuing 'bad' ones (e.g. an entrepreneur's profit). Clinical decision analysis, as we will shortly see, is simply a logical framework within which human beings can specify the quantifications and calculations necessary for identifying the course of action that is in the patient's best interest.

The six modes of cognitive continuum theory vary greatly in respect of how much control (i.e. ability to manipulate the variables in the situation) the judge or decision maker possesses. The modes vary greatly in respect of the overtness and visibility of the inquiry process and hence in their reproducibility (the key characteristic of the scientific experiment) and accountability (i.e. ability to provide an account of what has gone on). Most obvious of all to hard-pressed clinicians they vary enormously in their time and resource requirements. The greater the demand for reproducibility and accountability, and the greater the time and resources available, the greater will be the analytical content of judgement and decision making according to the descriptive hypothesis of cognitive continuum theory.

But while there are undoubted time and resource constraints on practitioners and while some aspects of task structure are virtually unchangeable, there does exist considerable freedom at the level of the individual practitioner — and much more at the level of the profession — to 'choose' the level of structuredness of tasks (including time pressure). If this is so it would surely follow that a profession's obligation is to choose that 'available' level of structuredness which will induce the mode of practice that will produce the best outcome. In making this choice, it will be vital, ethically, to ensure that judgements of 'unavailability' reflect genuine constraints, not simply professional preferences (as in the use of any simple 'acceptability to practitioners' criterion).

Decision analysis

A decision always involves choice among two or more options (e.g. 'remove' or 'leave') and denotes the cognitive preliminary to the action (or inaction) which is to follow now. It therefore embraces any type of 'wait and see' choice, selection of which is correctly seen as a decision rather than as postponement of a decision. There is no such thing as a postponed decision, only a decision to wait, for a specified or unspecified time, before a subsequent and new decision is taken.

Decision is to be contrasted with judgement in that decision involves choice among alternatives, whereas judgement involves assessment of alternatives — usually on some sort of scale, ranging from the most primitive binary and nominal one such as 'benign' versus 'malignant' to the most sophisticated and fine-grained multi-attributive one. So one judges the chances of surgical complications, and one judges the quality of life that will follow if they occur, but one decides whether to operate or not.

Genuine decision making involves situations where the possible outcomes of action, including 'doing nothing', are: (a) subject to uncertainty, and (b) of differing desirability. Any systematic exploration of a decision will necessarily involve the framing of the problem as a choice among alternative actions (e.g. operate or not), with each action leading into numerous possible scenarios. Each of these scenarios will reflect particular uncertainty resolutions (e.g. the appendix was or was not inflamed, the surgeon's scalpel did or did not slip) and the arrival at a particular outcome state (e.g. good health, moderate disability, severe disability, death). Clinical decision analysis provides the

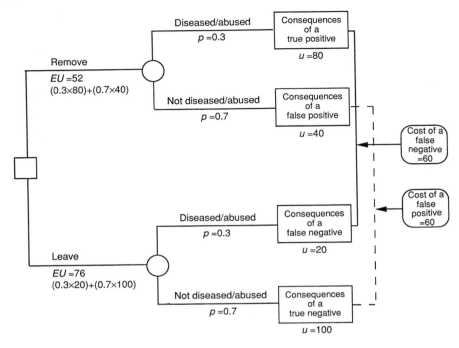

Figure 8.2   Basic decision tree: *p*, probability of disease/abuse; *u*, assigned utility value of the outcome; *EU*, expected utility of each choice (remove or leave), the sum of the utilities of its two outcomes (multiplied in each case by the probability of that outcome occurring). Here the optimal choice is to 'leave' (higher *EU* value).

framework within which the judgements of chances and the judgements of desirabilities necessary for optimal choice can be made independently — so as to avoid the lurking dangers of cross-contamination — and logically integrated.

But first the possible scenarios must be laid out explicitly in the form of a 'tree'. The simplest clinically relevant decision tree is found in Fig. 8.2.

This most basic of structures suggests that there are only two available options, in this illustration 'remove' and 'leave' (for example, a possibly diseased appendix or a possibly abused child.) It further suggests that there is uncertainty about whether or not the underlying condition (appendicitis, abuse) is present. Finally it suggests that there are four different outcome states, at least some of which will be regarded as more desirable than others. These four states are conceptually equivalent to the consequences that flow from making, or not making, the two sorts of error that are possible in this situation ('false positive' and 'false negative').

This tree has, in fact, also been provided with the numerical 'fruit' necessary to establish the optimal option — probabilities on the chance branches to quantify the uncertainties and utilities on the outcome nodes to quantify the relative undesirability of the outcome states. Given the numbers inserted here, and given acceptance of the principle of 'maximising expected utility' (selecting the option with the highest score when the desirability of outcomes is weighted by their chance of occurring), the optimal choice is to 'leave'. Other numbers might change this. Whether they will depends on the magnitude of the change and the 'sensitivity' of the decision structure to the change. As one illustration, in our tree a probability of disease/abuse higher than 50 per cent will make removal the optimal decision.

No further details of the technique can be provided here, but it should be emphasized that there are few limits to the complexity of decision trees. While most real-life cases are more complex than the basic one above, the limitations of decision analysis lie not in the technique, nor in the software that is desirable for its full quantitative implementation in most cases, but in the ability of human beings to think clearly about great complexity — and articulate it. It is common to find clinicians arguing that they cannot structure a problem as a tree (even with expert help) because of the complexity of the problem. Nor, they say, can they provide the necessary chance and desirability assessments. Despite this they express great confidence in their ability to make the decision successfully, or at least well.

Accidents and 'errors of judgement and decision'

So, in decision analysis there is a goal (which is to maximize the best interests of the patient, subject to constraints), actions (with choice needed between the alternative possible actions or 'options'), events (which, except in the case of tautologies, are chance ones in the sense that they have a probability between 0 and 1 and not either), and outcomes (which are states of the world at a moment of time or through a series of moments of time).

Is it possible that any outcome of a medical intervention chosen on the basis of a decision analysis could legitimately be called an 'accident'? This is our central question and to pursue it we need to see that in all medical decision making four main subtasks must be tackled somehow and to some degree, decision analysis being distinctive only in insisting that they be done separately and explicitly.

Task 1: structuring the decision by setting out the possible scenarios that will follow from adopting each of the available options. One of the key questions here is obviously to determine what options are 'available'. This can be viewed as a judgement task of a particular sort. What is judged to be an 'available' option for a particular patient or practitioner may be affected by clinical, religious, ethical, legal, economic, or other considerations. For example, if in our tree the 'removal' was that of a fetus incapable of independent life from the mother's womb, some people would argue that it is not an available option on the grounds that abortion is simply wrong. This sort of judgement will determine whether an action option gets on the choice menu and becomes a choice branch. The sorts of judgements involved in the next two tasks relate to the evaluation of the contending options.

Task 2: making a judgement as to the magnitude of the chances involved wherever there is uncertainty involved in a scenario. The uncertainty may relate to whether an event will or will not occur in the future (e.g. the abuse will continue if the child is left, the anaesthetist will or will not turn on the wrong tap if the operation is undertaken), or whether something is or is not the case at the moment (e.g. the appendix is or is not diseased).

Task 3: making a judgement as to the desirability of each possible outcome state (i.e. assessing the relative worth of states such as 'normal good health', 'confined to bed and unable to undertake self-care', 'dead'). Making this type of judgement may, of course, involve the making of internal 'subjudgements', for instance, of the relative importance (weighting) of health dimensions such as mobility, mental function, and pain.

Task 4: choosing and applying a principle for integrating the foregoing judgements, that is, for combining the uncertainty and desirability assessments pertaining to each available option into the *global* assessment for the option that is essential to its direct comparison with each of the other available options. In decision analysis this integrating principle is that of 'maximizing expected utility' (MEU) — weighting desirabilities (utilities) by their chances (probabilities) and selecting the option with the highest result. (Some readers will be aware that there is considerable dispute over the normative status of the MEU principle. The fact that the author does not find these arguments against MEU convincing is less important than noting that the explicit specification of chance and desirability judgements in the tree enables the effect of adopting of any alternative well-specified integration principle to be explored. The on-going debate about MEU is therefore not of any great significance for our general point regarding the implications of decision analysis for the 'medical accident'.)

From these task descriptions it should be evident that each of the three cognitive 'tests' for an 'accidental' diagnosis — unintentionality, unforeseeability, and harmfulness — are seriously affected once a decision analytic approach is adopted.

The intention of the medical intervention has been explicitly raised as soon the necessity of structuring the decision and adopting some integrating principle is acknowledged. And once the necessity of formally articulating intention is established, that of a medical encounter can hardly be said to be other than to select and undertake that course of action which is in the best interest of the patient, before the event — subject to constraints and uncertainty. It is meaningless to talk of any single outcome of an intervention flowing from a decision taken on the basis of this principle as being 'intentional' or 'unintentional' — whether this outcome is a 'harmful' one or not. Precision in the use of terms such as 'hope', 'intention', and 'expectation' is vital here, even if practitioners dealing with distressed patients find 'being economical with clarity' in their use very tempting. The hope is obviously to produce the best outcome, but the only intention can be to take and implement the best decision, which in decision analysis involves selecting the course of action with the highest expectation (in the precise mathematical sense defined above).

Moving from intentionality to harmfulness, the various possible outcomes of medical intervention almost always differ in their degree of desirability. As we have seen in our earlier tree this is something that decision analysis requires to be addressed explicitly and quantitatively.

Utility has a specific meaning in decision analysis and two aspects of this meaning are worth emphasizing. First, utility is a function of preferences and a utility measure will therefore reflect whatever is of value (or dysvalue) to somebody. It may therefore be very far from a 'utilitarian' assessment in the contemporary, everyday sense of that word, reflecting any altruistic and non-material values the person holds. Secondly, the fact that 'utility' seeks to capture the 'desirability' of an outcome state in a single number does not rule out an underlying multidimensional structure of values. The requirement for a single number simply registers and emphasizes the fact that the identification of what is 'best' will typically require tradeoffs to be established between different outcome dimensions — for example, between pain and mobility or between quantity and quality of life.

Compared with other types of decision aid, decision analysis has a unique advantage in that it provides — and indeed virtually insists on — the explicit incorporation of individual patient values and preferences. It is therefore the only one which is designed to aid decisions about individual patients as distinct from either aiding judgements about them (as do most data-based diagnostic aids or 'scoring systems') or, in seeking to aid decisions about them, assumes that their values and preferences are identical (as do most knowledge-based algorithms or expert systems). This is indeed the prime reason for concentrating on decision analysis in this chapter.

Finally, to 'foreseeability'. The key commandment of decision analysis is 'Thou shalt separate assessment of the probability of an outcome occurring from the assessment of its utility'. The assessment of the desirability of outcomes is a quite different task from the assessment of their chance of occurrence. One of the most likely reasons why inferior chance assessments are made is contamination by desirability considerations (as in, for example, 'wishful thinking'). As with utilities, decision analysis insists on the numerical representation of all uncertainties in the form of probabilities.

In many respects the key issue in relation to 'foreseeability' is whether or not a chance event (i.e. an uncertainty, such as whether the surgeon's knife will slip) should be represented in a decision tree. It cannot be stressed too strongly that this will almost never be a question of whether or not it is possible. Unless it is impossible by logical definition (not empirical judgement) then an event has some probability of occurring. The practical issue in decision analysis is accordingly whether its probability is 'low enough' to justify omitting it on the ground that it was 'unforeseeable'. This, by definition, is largely a question of specifying where a cutoff on the probability scale is to be drawn—at one in 10? one in 100? in 1000? in 10 000? in 100 000? in 1 000 000? in 10 000 000?

How should this cutoff for deciding whether a chance branch is to be included in the tree be set? The one clear part of the answer is that the probability threshold must be set independently of any utility considerations — precisely in order to ensure that probability and utility considerations are each given their proper weight in the analysis! If it is not set independently then double counting and confusion must result.

Failure to observe this independence leads one down the route followed by those (including legislators and judiciaries!) who talk in terms of the hybrid concepts of 'risk', 'danger', and 'safety' and the associated hybrid judgements of 'riskiness', 'dangerousness' and 'safeness'. All three sorts of hybrid judgement — when they are not correctly 'bred' from independent judgements of chance and desirability — are part of the linguistic swamp that produces 'accidents'. But if they *are* correctly bred from their distinct conceptual parents, as in decision analysis, their naïve attractiveness and apparent usefulness disappears. Why? Because the tree formulation establishes that the issue in medical encounters is not whether any particular option is 'dangerous' or 'safe', or involves 'risk'. (Adding apparently quantifying qualifiers like 'too', 'unacceptably', 'significant', 'serious', or 'substantial' to these terms merely compounds the confusion.) The issue is simply whether it is the one that is in the best interest of the patient or not.

Given its detailed and explicit attention to both probabilities and utilities — and to the principle for integrating them — it should be clear that all the substantive concerns evoked in these hybrid terms are adequately and properly addressed within the decision analytic formulation.

If there are no 'accidents' what are there?

'Will decision analysis let practitioners off the hook?', some (non-practitioners) may be asking. The quality of a decision made via decision analysis remains open to evaluation. In fact, because it opens up the conceptually distinct aspects of decision making — decision structuring, chance assessment, desirability assessment, and integration — for separate examination, it offers the prospect of an evaluation which is both diagnostically accurate (i.e. focuses on the appropriate source of pathology, if any) and therapeutically relevant (i.e. will assist in developing preventive measures).

There is no way that the conclusion of such an (after the event) evaluation of a decision made on the basis of a decision analysis could be that the outcome that occurred was 'an accident'. In the principal-agent setting of the medical encounter the 'explanation' of all outcomes of an intervention is properly seen as equivalent to the reasons for undertaking that intervention as opposed to alternatives, or none. And these reasons are constituted by the decision tree — its structure, the probabilities and utilities with which it was 'fruited', and the integrating principle which was used to resolve it, together with the reasons why all these aspects of the tree are as they are.

So, let us imagine that an undesired/unhoped for/harmful outcome of the sort that currently provokes the term 'medical accident' occurs, but following medical intervention based on a decision analysis. What are the possible attributions?

A key distinction is that between optimal analysis and optimal care.

Like any other outcome, however desired or undesired, the sort of undesired outcome that evokes the term 'accident' can only be either: (a) a chance one that followed from the provision of optimal care, that is, the selection and carrying out of the optimal course of action; or (b) one — almost always still a chance one — that followed from suboptimal care, that is, the selection and carrying out of a suboptimal course of action.

Note that even if the outcome occurred in the context of optimal care this does not necessarily mean that it was based on optimal (error-free) analysis. In principle there is a normatively best tree — best structure plus best 'fruit' — for every decision. By definition this tree will produce 'optimal care', that is, result in the selection and implementation of the course of action that is in the patient's best interest. Optimal analysis will guarantee optimal care. But the reverse is not true: optimal care does not necessarily imply optimal analysis. Any deviation of any of the elements of the tree from the 'best' has the potential to produce suboptimal care, but only the potential. Whether it will actually do so depends on the consequences, in the case concerned, of making an error(s) of one or more of the four possible types: error in decision structuring, error in probability judgement, error in utility specification, and error in decision integration.

The point is easiest to illustrate by reference to errors in the judgement of a probability or specification of a utility.

Let us assume the decision tree includes a 'test' option. If the test is positive one action will be taken, if it is negative a different one. If the best available assessment of the sensitivity (True Positive Rate) of the test in the relevant population is 70 per cent and no study shows a higher figure, then a doctor who used 95 per cent will be committing an error of probability judgement. However, this error will have no impact on the optimality of the decision unless the 'threshold value' for the TPR lies between 70 and 95. The threshold value is that value at which the optimal option 'flips'. Below it one option is optimal, above it another becomes optimal. So if the optimal action is chemotherapy when some specified probability is 70 per cent, but surgery when it is 95 per cent, then somewhere in between 70 and 95 there must be a probability at which the options are perfectly balanced and the decision is a 'toss-up'. If this threshold value is 90 per cent, then a doctor may make a considerable error of probability judgement — using a figure as high as 89 per cent and as low as he likes — without this resulting in suboptimal care. (Recall that the best available assessment of the 'true probability' is 70 per cent.)

The implication is, of course, that practitioners need to be aware of where the threshold values lie, because a numerically small error of one probability (or utility) judgement can lead to a suboptimal decision while a numerically larger one of another may not.

It is vital to note that the simple term 'error of judgement' is (like the term 'accident') diagnostically useless, in that it does not distinguish, and implicitly confounds, the totally distinct types of cognitive task involved in decision making by decision analysis. By contrast 'error in decision structuring', 'error in probability judgment', 'error in utility specification', and 'error in decision integration' are specific diagnoses and target precisely what is needed for remedial and preventive work.

Taken in conjunction with the attribution to 'optimal care', these four errors generate five alternative and well-specified attributions for the outcome. The requirement for explicitness before the event in decision analysis means that there should be much less difficulty establishing (after the event) 'what happened' or rather why (at a decisional level) what happened did happen. The arrival of decision analysis will therefore in no way 'let doctors off the hook', but simply provide a much fairer and more systematic, open, and accountable way of establishing what, if anything, 'went wrong' and of furthering the most appropriate assignment of responsibility.

Because decision analysis makes it clear that optimal judgement and decision making is a task involving four very different subtasks, it also makes it evident that skill and ability in any one of these does not necessarily confer any expertise in any other. So a doctor who is excellent at establishing the available management options may or may not be good at assessing diagnostic or prognostic probabilities, or at eliciting the patient's utilities.

Failures in performance

There is almost certainly at least one aspect of the above argument that will be puzzling many readers. Where do failures in performance — in carrying out the optimal course of action as identified by decision analysis — fit? Are not they the most frequent cause of 'accidents'?

The most important consequence of adopting a decision analytic approach is that the distinction between errors of judgement and decision making and errors in action and performance loses its attributive significance, because the latter are subsumed in the former. How? If a surgeon's knife slips and he or she cuts the wrong nerve then the only relevant attribution is to the judgement concerning the probability of this occurring. Did the decision to operate incorporate the 'best' assessment of the chance of this unhappy event happening? If the surgeon had said there was no chance of this happening — and it did — then he or she would be absolutely guilty: the one indefensible error is to give a probability of zero/100 per cent to something which does/does not occur. If he or she had said there was a 1/1000, or some other specific chance of it occurring, then the attributional issue is solely concerned with the goodness of this assessment and not with why the knife slipped. Investigation of why the knife slipped on this particular occasion, while fascinating and capable of absorbing vast resources (personal, medical, and legal), will not necessarily, even if successful, throw any light at all on whether the best assessment of the probability of slipping was incorporated in the decision analysis.

Legal standards: optimal care and the 'reasonable tree'

We have talked throughout of the 'best tree' and 'best fruit'. Surely the legal standard is always going to be, as it is now, what is 'reasonable'?

It probably will, but when a decision analytic formulation has been adopted the discussion will be about the reasonableness of the tree as a whole and not about the reasonableness of any single course of action as such. It will not be a question of getting some colleagues to say 'in the circumstances I think option X, which was what the defendant did, was a reasonable thing to do'. The decision has been 'decomposed' and it will therefore be the reasonableness of each component of the tree — structure, probabilities, utilities, and integrating principle — and of each element of each component that will need to be assessed.

And at the component level the issue cannot be whether a particular judgement was somehow reasonable 'in itself', but only whether it was a reasonable estimate of the best. For example, it cannot be a question of whether a probability (or utility) assessment is reasonable, *per se*, but only whether or not it is a reasonable estimate of the current best assessment for this case.

Since decision analysis is a mode 4 technique, it will be perverse if the legal arguments concerning what is and is not 'reasonable' at the component level are allowed to remain at mode 5 if there is mode 4 evidence available. It should be little use for a witness to come to court and say 'in my experience the probability is 95 per cent, the very one that my colleague, the defendant, has used' — when the best available assessment for the relevant population in the literature is 70 per cent. This, of course, is precisely the sort of testimony that does count when the standard for reasonableness is the mode 5 one of what a 'respectable body of professionals' — not necessarily a majority — do or say.

Conclusion

The world of medicine may never adopt decision analysis (or other mode 4 techniques and systems) on any significant scale and in that case this chapter will remain entirely hypothetical, rather than merely futuristic. Whatever the reason for the continuance of mode 5 and 6 judgement and decision making — professional power, patient distaste, or both — we will then be left with the current situation concerning 'medical accidents', which is, to put it mildly, a conceptual mess. Those attempting to clear up afterwards are entitled to everyone's best wishes, but I urge them to consider seriously the suggestion that the mess is inherent in the current modes of practice. Legislation and/or judicial changes may well be helpful, but only if they change those modes of practice rather than trying to deal with their consequences.

It need hardly be said that nothing in this chapter is intended to suggest that those who suffer harmful outcomes following medical intervention (including intervention based on decision analysis) should not receive compensation. But it does imply, it is suggested, that the issues of who should be compensated, and by what mechanism, would be much better tackled when the decisional basis of medical intervention is fully exposed and established — as it is when decision analysis is undertaken.

References

Böckenholt, U. and Weber, E. U. (1992). Use of formal methods in medical decision making: a survey and analysis. *Medical Decision Making* **12**, 298–306.

Dowie, J. (1992). *Professional judgment and decision making: introductory texts* (3rd edn). The Open University, Milton Keynes.

Dowie, J. (1993a). Decision analysis: the ethical approach to medical decision making. In *Principles of health care ethics* (ed. R. Gillon). Wiley, Chichester (in press).

Dowie, J. (1993b). Clinical decision analysis: introduction and background. In *Analysing how we reach clinical decisions* (ed. A. Hopkins and H. Llewellyn). Royal College of Physicians, London (in press).

Hamm, R. M. (1988). Clinical intuition and clinical analysis: expertise and the cognitive continuum. In *Professional judgment: a reader in clinical decision making* (ed. J. Dowie and A. Elstein), pp. 78–105. Cambridge University Press.

Thornton, J. G., Lilford, R. J., and Johnson, N. (1992). Decision analysis in medicine. *British Medical Journal* **304**, 1099–1103.

9

Stress, psychological problems, and clinical performance

JENNY FIRTH-COZENS

There are numerous reports, largely from North America and Britain, that symptoms of stress and psychological problems are particularly high in members of the medical profession. This chapter will review the evidence for this, and look at the ways in which these factors might affect patient care, including accidents.

Psychological problems in medical practitioners

Until very recently research in this field was distinguished from studies of stress levels in other occupational groups by the fact that the measures used when considering the psychological symptoms of doctors were, and often still are, rather more clinical and diagnostic than those simply describing stress or strain (Firth-Cozens, 1987a). For example, studies look for psychopathology such as schizophrenia (Heath *et al.*, 1958), depression (Reuben, 1985), or alcoholism (Sclare, 1979), rather than for levels of 'occupational stress' which are reported for other occupational groups (for example teachers (Kinnunen and Leskinen, 1989) or public transport operators (Carrere *et al.*, 1991)), or for the even more general attribute of negative affectivity (Depue and Monroe, 1986) which is increasingly being used to sum up the highly correlated factors of job dissatisfaction, stress, and neuroticism (Ormel, 1983).

This clinical emphasis has the effect of making medical problems seem rather more dramatic than those experienced in other jobs, and also of making it difficult to compare doctors' work-related symptoms with others. On the other hand, the narrower grouping of symptoms implied by diagnosis is likely to be more useful in terms of predicting the types of performance deficits which might arise as a result of disorders such as depression.

DEPRESSION

The emphasis on depression (as opposed to some other diagnoses) in medical practitioners is by no means arbitrary. Rucinski and Cybulska (1985), in a British review of psychiatric illness in doctors generally, concluded that, although reliable data about this was difficult to obtain, the most common diagnoses remain affective disorders (which account for 21–64 per cent of doctors' admissions to psychiatric hospitals, depending on the study), and alcoholism and drug addiction (which account for between 51–57 per cent). Studies which consider the suicide rates for doctors also confirm that depression is a relatively high risk factor for them (Steppacher and Mausner, 1974; Pitts *et al.*, 1961; Rose and Rosow, 1973), especially in women doctors (Firth-Cozens, 1990).

Studies of depression in medical practitioners have used a variety of psychiatric interviews and questionnaires to arrive at their diagnoses, but have usually concluded that a far larger proportion of doctors suffer from depression than is found in the general population. Most of these studies have been concerned with junior doctors, in particular the first postgraduate year. For example, Valko and Clayton (1975), using an interview schedule, found that 30 per cent of 53 newly qualified doctors had suffered from clinical depression. Similarly, Reuben (1985) used the Centre for Epidemiological Studies Depression scale (CESDS) monthly during the first three years after graduation. He found a large peak between three and six months into the first year with up to 38 per cent having scores within the depressed range at that time. When classified by year of training overall, 'depressed' responses occurred in 29 per cent of first-year, 22 per cent of second-year, and 10 per cent of third-year postgraduates. This decrease suggests that factors of the job which might be contributing towards depression (such as long hours and reduction of social support), or of the individual (such as feelings of inadequacy or overwhelming responsibility) may reduce with time. However, these scores remained consistently high over the three years for doctors on rotations such as in the Intensive Care Unit where sleep deprivation continues to be high, and where death and dealings with distressed relatives are frequent events.

The same measure was used in a larger study by Hsu and Marshall (1987), who looked at scores from interns, residents, and fellows at a number of hospitals. Overall, they found 415 (23 per cent) of the 1805 respondents showing some degree of depression, with 72 of these categorized as severely depressed. As in Reuben's study, the first postgraduate year had the highest proportion of depressed respondents (31.2 per cent). The CESDS used in these studies finds an average of 15 per cent depression in a community sample, indicating that junior doctors have substantially higher levels of morbidity than the general population, despite the fact that, as professionals, they might be expected to have lower levels.

In the UK in my own study of 171 young doctors followed up from their fourth year as medical students (Firth-Cozens, 1987b; Firth-Cozens and Morrison, 1989), I used the depression scale of the SCL-90 (Derogatis *et al.*, 1973) and found that 28 per cent of doctors in their first postgraduate year scored at a level indicative of clinical depression. The cut-off score was derived from norms provided by a group of male and female professional workers who had all met a criterion of depression through Present State Examination interviews. Only one of these junior doctors was taking anti-depressant medication, suggesting that few if any had sought help for their problems. Again this proportion is much higher than that found in the general community; for example, Bebbington *et al.* (1981), in a British sample of 800 Londoners, found one–month prevalence rates using the Present State Examination of 6.1 per cent for men and 14.9 per cent for women.

Although those most studied have been in the junior years of medicine, all the studies agree that depression rates in medical practitioners as a whole are above those in the general population, especially when comparison is made with other professional groups. Since the manifestations of clinical depression — poor decision-making, memory and concentration deficits, selective attention, and interpersonal problems — will all have negative implications in terms of patient care, it seems clear that the emotional state of doctors is a cause of particular concern.

ALCOHOL AND DRUG ABUSE

Alcohol and drug abuse have clear links with depression (Goodwin, 1985), and clinical experience tells us that individuals who are experiencing difficult times — especially psychologically difficult — will increase their alcohol intake. Whereas everyone has ready access to alcohol, access to drugs is available to very few: primarily to health workers, but especially to doctors since they have the possibility of self-prescription. It is not surprising, therefore, that some reports both in the USA (Vaillant *et al.*, 1970), and in Britain (Sclare, 1979; Brooke *et al.*, 1991) have expressed concern about the high levels of alcohol and drug abuse within the medical profession. Although mention is usually made of the risks this might place on patient care, most studies show an understandable concern for the doctors themselves. For example, 20 per cent of physician suicides in the USA are reported to be associated with drug abuse and 40 per cent with alcohol abuse. In a British study which followed up 36 of an original 41 alcoholic doctors who had been hospitalized at the Maudsley in London 5 years earlier, only 8 were practising medicine at an adequate level, 9 were still engaged in addictive drinking, and 5 had died from cirrhosis or suicide (Murray, 1976a). Murray (1976b) concluded that these casualties are represented proportionally more in doctors than other professionals; he studied first admission rates to Scottish psychiatric inpatient beds of male doctors compared with other social class I males with a primary diagnosis of alcoholism, and found them to be 2.7 times higher. In addition, the liver cirrhosis mortality rate in

England and Wales for doctors is 3.5 times that of the general population. On this basis, Glatt (1976) estimated that there were 2–3 thousand alcoholic doctors in this country at that time.

More recently, in the USA, McAuliffe *et al.* (1991) conducted a random survey of doctors' and medical students' alcohol use and abuse and concluded that they were no more vulnerable to abuse than pharmacists or other professionals. The average number of days per month when alcohol was used was 10.5 for men and 6.5 for women, and 4 per cent of both doctors and of the comparison group of pharmacists reported having a drinking problem at some time since leaving college. However, four doctors in every hundred describing alcohol abuse is certainly not low, especially when self-report is likely to show underestimation of the problem. In addition, they found that, whereas alcohol use decreases significantly with age in other professional and general groups, in the medical profession it increases at the same rate.

A recent paper by Brooke *et al.* (1991) agrees that the problem should not be underestimated. Their trawl of psychiatric notes for two NHS (public) hospitals over 20 years revealed 144 doctors with admissions for drug and alcohol problems. Such a number is obviously a very small proportion of the actual total since many never receive psychiatric help or, if they do, most seek private treatment. Others present with physical diagnoses such as cirrhosis. None of their sample had left medicine permanently, though 22 per cent were unemployed at the time of treatment. Of equal concern in terms of patient care was the statistic that the average duration of the problem prior to seeking help was 6.7 years for alcohol abuse and 6.4 for drug misuse. Again, reflecting the US findings suggesting an increase of problems with age, the authors found that the mean age that alcohol became a problem was 37, drugs 34, with intravenous use from 32. The range was 18–59 years, indicating that doctors are entering the profession with the problem as well as beginning it in their senior years. However, the older mean age is of concern, since the much larger autonomy of senior doctors means that problems are less likely to be reported, and their decision-making holds more power.

My own sample of 171 junior doctors (Firth-Cozens, 1987b), like those of McAuliffe *et al.* (1991), showed 4 per cent describing their alcohol use as heavy and frequent with another 17 per cent describing it as heavy and occasional. There was a correlation of 0.37 ($p<0.001$) between the two testings showing that on the whole a pattern of drinking behaviour was being maintained. In addition, 7 per cent of junior doctors reported that they were using drugs 'recreationally'. Since the sample was working a mean of over 90 hours a week, there must inevitably have been interference with the work of a substantial proportion due to drug or alcohol effects.

There was no correlation between alcohol use and stress levels at either time of testing, but it may be that for some subjects, especially males, alcohol in the short term is a successful way to maintain psychological well-being. In an unpublished study of 724 junior anaesthetists, there was a suggestion that alcohol was being used in this way. For example, one respondent, asked the way that 'stress-related symptoms (e.g., tiredness, tension, depression, alcohol, or drug consumption) have affected your patient care' answered: 'I sometimes have an alcoholic binge after the death of an ITU patient which hurts particularly.' In terms of the links between long hours and alcohol use, another respondent said: 'I think there is a general trend in juniors to do little exercise, even when they have time off, as they are so tired, so socialising with alcohol is much easier.'

In this study there were mentions of the ways that alcohol affects patient care; for example: 'I am so tired so often that I just want to collapse in a chair and have a few drinks. This causes shaky hands the next day so epidurals etc. are more difficult'. This chapter will not review the vast evidence that alcohol affects physical, cognitive, and interpersonal performance. Most people would see this as long accepted. What is important is that at least four doctors in every hundred are admitting to drink problems, that some are using drink as a coping mechanism rather than learning more appropriate ones, that alcohol problems exist for many years before discovery or change, and that alcohol use increases with age. To illustrate, our young doctor with the shaky hands reported above, as well as the one using alcohol to blot out sadness and possibly self-blame, may eventually have the time and good fortune to find a more appropriate way of tackling such job-related problems. On the other hand, it may be more likely that the pressure of their careers over the following years and the difficulty that many doctors experience in seeking help (see, for example, 'Dr Magoo' in Mandell and Spiro, 1987) will lead them to increase their alcohol intake, not just because they need more to gain the same effects and to 'cure' the shaky hands but also to pre-empt the sadness caused by patient death.

LEVELS OF STRESS

Little research has been conducted on general levels of stress in more senior practitioners, but self-reports show that the symptoms of stress are high amongst juniors doctors. For example, using the General Health Questionnaire, I found that half of a population of junior house officers was reporting symptom levels indicative of emotional distress or 'caseness' (Firth-Cozens, 1987b). The correlation between GHQ scores and depression was 0.73 ($p.<0.000$). A recent paper by Brazier *et al.* (as yet unpublished) used the same instrument and looked at all levels of house officer, and confirmed this high proportion, showing 46 per cent to be above the cut-off level. The usefulness of using a well-validated measure of general health or distress such as this is that populations can more easily be compared with other occupational groups. In fact, health workers generally show high proportions (Jones, 1987;

West *et al.*, 1988) of between a quarter and a third. By comparison, a study of a group of young employed workers and working men reported 8.3 per cent and 8.9 per cent respectively (Banks *et al.*, 1980).

It is clear that the levels of stress, depression, and addiction in doctors are higher than in the general population, and probably higher than other professional groups. While the consequences for patient care of alcohol and drug abuse are obvious, those of depression and stress are less so, and their relationship to sleep deprivation is more complex.

Factors which affect job performance

The effects of sleep loss and sustained pressure upon performance have been studied in other occupational groups, but less so in doctors. Whether these effects act directly upon performance or do so via the individual responses of stress or depression is much less well understood (Motowidlo *et al.* 1986; Jex *et al.*, 1991). This section will consider the extent to which sleep loss and overwork are seen as causing stress, and then go on to look at the effects of these three factors upon performance.

SLEEP LOSS, OVERWORK, AND SYMPTOMS OF DISTRESS

McCue (1985), commenting on the problems of young doctors, considers that 'time pressures and sleep deprivation constitute the major stresses of residency training, adversely affecting the ability of residents to learn, the quality of medical care they deliver, and their ability to respond appropriately to urgent problems.' Certainly when asked about the sources of stress (for example, by scoring stressfulness levels against a list of potential sources), junior doctors invariably score overwork and sleep loss as the principal causes of their difficulties (Ford, 1983; Small, 1981; Hurwitz *et al.*, 1987; Firth-Cozens, 1987b).

Nevertheless, the causes may not be so clear-cut. For example, a study by Clark *et al.* (1984), looking at predictors of depression during the first postgraduate year, found no significant relationship between perceived or actual workload and depressive symptoms. Similarly my study (Firth-Cozens, 1987b) showed no relationship between the number of hours worked and either stress levels or depression levels, though this may have been due to a fairly narrow distribution of hours worked which could have limited the possibility of significant correlations. In terms of their individual responses to such difficult work conditions, while 25 per cent were above threshold (cases) when house officers but below threshold while students (which is the direction one would expect if aspects of the job alone were creating the stress), 16 per cent actually changed from being cases as students to being non-cases as doctors, providing some evidence for individual responses to long hours and little sleep.

More evidence for the individual responses to their jobs come from a longitudinal regression study of my sample (Firth-Cozens, 1992a) which looked at the causes of stress and job attitudes. It found that no objective job-related measures (hours slept and worked, number of beds) predicted stress or depression levels; instead, the most significant predictor was the level of self-criticism measured while a student, which in turn was predicted by perceived early family relationships. When subjects were asked to complete a qualitative account of a recent stressful event (Firth-Cozens and Morrison, 1989), accounts reflecting overwork represented 11 per cent of the total, whereas 'dealing with death and dying' concerned 19 per cent followed by 'relationships with senior doctors' (16 per cent) and 'making mistakes' (13 per cent). This may indicate that overwork is not such an important factor in inducing stress as other aspects which may for some individuals produce higher levels of anxiety which are denied in favour of 'overwork' in more traditional questionnaires (Mumford, 1983). On the other hand, it may be that the tiredness from overwork makes the experience of patients' death or dying or of relationships more difficult to take. Nevertheless, the main predictor of difficult relationships with consultants was not hours worked or amount of sleep, as might be expected, but being in a teaching hospital and reporting a difficult relationship with one's father (Firth-Cozens, 1992a), again emphasizing the individual differences of reactions to work factors.

Although the evidence concerning the effects of sustained work on stress and depression is ambiguous, the effects of sleep loss on mood generally shows rather more agreement. For example, Friedman *et al.* (1971), studying 14 interns, found that, when sleep-deprived, they felt significantly more sadness, and reported increased irritability, inappropriate affect, and depersonalization, as well as showing a performance defect in their ability to recognize abnormalities in medical data. Poulton *et al.* (1978), studying 30 British junior house officers in various degrees of sleep deprivation, found physiological signs of stress such as tachycardia, extrasystoles, or slight tremors, while some were kept awake with anxieties relating to patients and others by nightmares. Similarly, Hurwitz *et al.* (1987), using the Middlesex Hospital Questionnaire to identify psychological disturbance (or what they called 'demoralization'), found the best predictors of this to be an interaction between sleep deprivation and social deprivation caused by long work hours. They also reported some individual vulnerability, but did not say how this was measured. Again, in a study by McManus *et al.* (1977), 14 per cent of the 50 per cent of junior house officers who replied felt that their psychological well–being had been affected negatively by a lack of sleep. Further evidence for the effects of sleep loss on mood came from my own study (Firth-Cozens, 1987b) where there was a small but significant relationship between the number of hours of sleep in the preceding 48 and stress levels, and between difficulty of adjusting to sleep patterns and both stress and depression. However, both these problems will to some extent be affected by depression, so the directions of causality is somewhat ambiguous, and in regression analyses, the hours slept predicted no aspect of stress or depression.

While some studies show the effects of sleep loss on feelings of depression and symptoms of stress, others have found that it results in anger and hostility for some subjects (Ford and Wentz, 1986; Uliana *et al.*, 1984). Although this may at first seem contrary to those who find depression to be paramount, psychoanalytic theories see depression as being the inward expression of anger, and so, in any small group of doctors, individual differences of reaction may well predispose some predominantly to anger or to depression. Whereas depression may predispose people to cognitive mistakes, anger and hostility may reduce their levels of good interactions with patients. This will be discussed more fully later.

From these studies as a whole, and from some of the non-medical studies of sleep deprivation on performance which are described next, it is clear that, at least for a large proportion of individuals, mood deteriorates in various ways over a period without sleep.

THE EFFECTS OF SLEEP LOSS AND SUSTAINED WORK ON PERFORMANCE

In 1890, when the 84-hour working week at blast furnaces was reduced, accidents and absenteeism declined and productivity in some cases increased (Alluisi and Morgan, 1982). Certainly most studies which look at cognitive performance, such as vigilance and attention, do generally support the everyday presumption that working long hours causes deficits in the accuracy of tasks. For example, there is progressively less efficiency in detection of visual or auditory signals over time, after only half an hour of attention (Mackie, 1977; Warm, 1984), though environmental and individual factors will affect the rate of deterioration. Angus and Heslegrave (1985) followed subjects over 54 hours of continuous work involving information processing and found that reaction times, logical reasoning, vigilance, and other cognitive factors deteriorated after 18 hours to a maximum of 40 per cent of the individual's baseline efficiency, with a concurrent lowering of mood and motivation. These deficits were at their worst during the early hours of the morning (3 a.m. to 6 a.m.) when the circadian rhythm is at its lowest. Vigilance deficits could be critical in certain specialties such as anaesthetics and intensive care, but the wide array of cognitive decrements will affect most doctors' tasks to some extent.

Many of the junior anaesthetists in my unpublished study supported the view that tiredness due to sleep loss and job overload was directly responsible for deficits in their patient care. Some of these related to cognitive deficits causing poor standards of care, mistakes and even death. For example:

Tiredness from many hours of continuous work or after frequent calls during the night causes a definite increase in the number of attempts before successful venepuncture or arterial cannulation is accomplished. I am also slower to react to changes during anaesthesia requiring intervention.

Tiredness makes me unwilling to treat patients in intensive care as actively at night, leaving 'proper' treatment till morning.

I know I have a slovenly approach to ITU management when tired with marked reluctance to alter management.

Due to tiredness I forgot to remove a throat-pack from a dental patient. He became hypoxic in the recovery room and had a cardiac arrest. Fortunately he survived with no sequelae.

I gave a drug via an incorrect route after a long operation. I would put this down to lack of concentration from tiredness, after having worked solo all day. The mistake was spotted by myself within seconds and so no ill effects, but it was potentially very dangerous.

Under stress to keep up with demands for emergency anaesthesia I have on a couple of occasions inadequately assessed patients preoperatively and they have turned out to be much sicker than expected, leading to problems during anaesthesia that were probably avoidable and contributed to their demise.

Other accounts emphasized that tiredness caused mood changes of irritability and hostility which affected patient care and staff relations. For example:

Tiredness after being on call leaves me irritable, selfish, tendency to become short with colleagues and patients. As a result I become frustrated and even more grumpy because I know I would perform better if I had proper rest.

Tiredness, especially late at night often causes me to be short or rude or at least not as open and friendly to people as I would normally like to be.

When tired I tend to give much less explanation of what is going on to patients and staff. I worry more about it afterwards when I think how much better I could have behaved.

Most studies agree with these doctors that sleep loss, the inevitable consequence of long hours of work, adds to the deficits caused by sustained effort (Krueger, 1989). A number of studies have shown progressive deterioration in performance over a continual 72 hours without sleep. For example, Babkoff *et al.* (1985) found that subjects completed their tasks increasingly slowly, and this was accompanied by deteriorating mood and motivation, while Williams *et al.* (1959) found errors increased as well as slowness as sleep loss progressed. It would be comforting to think that people can become used to sleep deprivation and need less sleep over time, but the evidence suggests that this is not so (Webb and Levy, 1984). Most of these studies have been conducted in laboratory settings, but there is no reason why the results should differ in applied settings such as medicine.

Nevertheless, what is apparent, both in these studies and especially in those which have considered sleep loss and performance in doctors, is the extent to which individuals can rally themselves to perform well even after extremely long periods without sleep. It seems that the rate of decline in performance has as much to do with the intrinsic interest of the task and with factors in the individual as it does with a lack of sleep; for example, subjects improve in cognitive tasks if they are told that they can have a short sleep soon (Haslam, 1985). This capacity to rally in certain circumstances may well account for the very mixed results which have been found when junior doctors have been followed over periods without sleep. In these studies, the general methodology is to interrupt doctors during their shifts and to get them to complete a variety of cognitive tests appropriate to their jobs (for example, decision-making, vigilance, and learning), and sometimes also to measure mood.

The study by Friedman *et al.* (1971), described earlier, used tests of scanning for episodes of arrhythmia on an ECG tape, the accuracy of which deteriorated during the 32 hour period they considered. The task, although clinically relevant, was monotonous, and it has been suggested (Kjellberg, 1977) that it is these tasks, rather than those which might increase arousal, which will be most vulnerable to the effects of sleep loss. This was tested in the study by Poulton *et al.* (1978), described above, which used both grammatical reasoning tests and, for half the subjects, a brief laboratory report checking test. There were deficits in those doing solely the reasoning test after a sleep debt of only three hours. The other group was given the results of their reasoning test, compared to norms, before going on to do the report-checking. Their results showed that impairment came after eight hours sleep debt. This may have been due to the effects of knowledge of results or because the second task was more relevant and interesting and so perhaps increased arousal.

Another UK study by Deary and Tait (1987) followed 12 house officers, testing them cognitively after a night off duty, a night on call with no new admissions, and a night with new admissions plus calls. Performance on a memory test was significantly impaired after the third condition, though tests which were more related to their jobs showed no significant impairment. They noted, however, that some doctors performed better when they were tired than colleagues did when they were alert. This again emphasizes the important role of individual differences in these studies. It may be that the intervening factor between sleep loss and performance is mood state, which is affected differentially by little sleep or overwork, and/or it may be a personality characteristic of competitiveness or adaptability which allows these doctors to rise to the occasion of testing (or perhaps of responding effectively to a patient) in an alert fashion whatever their state of sleep deprivation. Certainly the results of sleep loss and performance in junior doctors are less conclusive than those conducted in laboratories; for example, in a study by Ford and Wentz (1984) there were no significant decrements in performance as sleep loss increased, even up to 72 hours.

In summary, there is considerable evidence, from experimental studies in particular, that sleep loss and sustained work have direct effects upon the types of tasks which make up the role of doctors, in particular junior doctors and those in specialties where tasks require continuous attention. However, there is also evidence that this may depend upon individual differences, and it may be that these job-related potential stressors are also acting indirectly, being mediated in their effects via mood; in particular, via depression and possibility hostility.

THE EFFECTS OF STRESS AND MOOD ON PERFORMANCE

Lazarus *et al.* (1952) saw stress as an intervening variable, with preceding causes or stressors, and behavioral consequences. A number of models have been postulated to explain this sequence. For example, Cohen (1980) suggested that stressors create conditions of information overload which causes cognitive fatigue, thus reducing the energy needed for the task. In particular, his review showed that a wide range of stressors such as overload, noise, and bureaucracy not only result in people performing less well in terms of accuracy, but also in 'a decrease in helping, a decrease in the recognition of individual differences and an increase in aggression.' Many of the anaesthetists' accounts of the results of tiredness in interactions with patients, described in the previous section, support this statement. Some researchers, regarding job dissatisfaction as virtually synonymous with stress, have used this as a mediating variable in similar models, again showing significant relationships between job satisfaction and various positive interpersonal behaviours, such as being helpful, co-operative, and listening to and showing consideration for others (Motowidlo, 1984; Bateman and Organ, 1983; Schneider and Bowman, 1985).

Motowidlo *et al.* (1986) attempted to test out their particular model which presumed that both job characteristics and the individual would contribute to the frequency of stressful events and that the individuals' characteristics (such as job experience, type A personality, and a fear of negative evaluation) would contribute to the intensity with which these events would be experienced. The frequency and intensity could lead to subjective feelings of stress and negative affect such as anxiety, hostility, and depression, which in turn would lead to performance decrements of inaccuracy, low tolerance of frustration, and interpersonal sensitivity. The subjects were 230 nurses, and their performance was rated for 107 of them by a supervisor or co-worker.

Using path analysis, they found that overload, unco-operative patients, criticism, and various interpersonal difficulties with colleagues (nurses and physicians) were related to feelings of stress, the perceived intensity and frequency of which co-varied with depression, anxiety, and stress levels, as did both cognitive and interpersonal performance measures. The frequency of events depended in part on where they worked; for example, those in medical and surgical units reported higher frequencies. Intensity of events depended at least in part on how much they feared evaluation, as did depression and anxiety. In terms of the final part of the model, the authors conclude that 'feelings of job-related stress lead to feelings of depression that cause nurses to perform less effectively in the interpersonal and cognitive/motivational aspects of their job'. Interestingly, feelings of hostility did not predict interpersonal performance, as was hypothesized, but depression had a significant negative effect on all performance variables, both cognitive and interpersonal (quality of patient care, interpersonal effectiveness, warmth and tolerance towards nurses and doctors), except tolerance with patients.

The role of psychological strain as a mediator between stressors and performance was also confirmed in a population of resident physicians by Jex *et al.* (1991). Performance here was measured subjectively by questions about unexplained absences, mistakes, missing deadlines, and conflict with co-workers. Abusive patients and changing schedules were the significant stressors in the regression analysis, while sleep deprivation/excessive hours were relatively unimportant. Girard and Hickam (1991), considering the predictors of clinical performance among internal medicine residents, again found that psychological state, when combined with various personal characteristics such as being older, predicted clinical ranking, and examination scores.

In my own study (Firth-Cozens and Morrison, 1989) a different methodology again suggested links between stress and poor performance. Describing incidents of stress which had occurred within the past month, 13 per cent (21) of doctors wrote about events which involved their personally making mistakes. These often ended with statements such as 'luckily the lady did not die' or 'we were thankfully able to save his leg'. No one in this study reported feeling responsible for the death of a patient (as they did in a previous study of intern's mistakes by Mizrahi (1984), and in my own study of anaesthetists reported earlier). Nevertheless, the mistakes described were still serious; for example:

I missed the diagnosis of pulmonary embolism and treated the patient as a case of severe pneumonia until the day after. The patient's condition deteriorated and only then was the diagnosis put right. I felt guilty and lost confidence.

and

Missing a diagnosis of perforated peptic ulcer in a patient — at least she is now well and survived. It made me feel useless at my job though.

The proportion of accounts concerning mistakes was also large in Mizrahi's study (1984), where he asked 'What were your most memorable experiences during training?', which elicited 21 per cent of descriptions referring to actual or potential mistakes. In addition, he found that serious and even fatal mistakes were made by half of the new interns interviewed in his study in the first two months of their jobs. His interview findings and my own questionnaire results were not anonymous, so they may reflect a bias against the reporting of mistakes.

We went on to examine the differences in the reporting of stressful incidents between those who had been above threshold on the GHQ on both occasions of testing (chronically stressed) with those who were below on both occasions (never stressed). The only significant difference in the types of stressors reported between the two groups was that the chronically stressed reported considerably more incidents of mistakes than those never stressed (23 per cent compared with 7 per cent). The preceding review makes it quite likely that there may have been actual differences in the patient care of the two groups; that is, that stressed or depressed mood was actually leading to more mistakes. However, it is also true that the chronically stressed were significantly more self-critical (Firth-Cozens, 1987b) and were more likely to have tried to cope with the event by dismissing it (14 per cent compared with 2 per cent). It may be that these individuals are simply blaming themselves for acts for which another doctor may blame others. On the other hand, these findings and some of the words used in the descriptions suggest that self-blame may have continued to fuel the depression which in turn may go on to affect performance in a vicious circle (this is also apparent from the anaesthetists' accounts quoted earlier). That the experience of guilt over mistakes continues for years in some doctors has been remarked upon by Mizrahi (1984) and by Margison (1987).

So far, we have considered the experiences of individual doctors who are feeling stressed, depressed, or hostile. However, whole organizations can also be described as stressed, sick, or neurotic. Merry and Brown (1987) describe characteristics similar to those we might find in a stressed individual: low self-image, low energy, poor communication, disagreements about goals, neglect of facilities, decreasing outputs, and recurring intensifying periods of crisis. In a fascinating series of studies, Jones *et al.* (1988) considered stress and medical malpractice at this organizational level, using 67 hospitals and more than 12 000 workers. In the first of two studies, they found that hospital departments with a current record of malpractice reported higher levels of on-the-job stress than did matched, low-risk departments, and in the second study they found that the stress levels of 61 hospitals correlated significantly with frequency of malpractice claims. However, it is also possible that some areas have overly litigious patients and that the experience of malpractice suits increased stress levels (Scheiber, 1987). The third study attended to these points with a longitudinal design which showed the effectiveness of a stress management programme for the whole of one hospital in terms of reducing the frequency of reported medication errors. These fell from a monthly mean of 10.3 in the year prior to the intervention, to 5.1 in the subsequent year.

The fourth study was a two-year investigation comparing the frequency of malpractice claims following the intervention. They found that the 22 hospitals which had used the programme (having been invited to do so by the authors) at the end of the two years had significantly fewer claims than the matched sample which did not. The hospitals were matched on bed numbers, frequency of malpractice claims during the preceding year, and whether they were urban or small town since urban patients are more likely to sue (Crane, 1987). In 1985 the experimental group had 31 claims compared to 9 in 1986 after the intervention, while the control group had 36 in 1985 and 35 in the following year.

Of course, there may have been something of a Hawthorne effect concerned with the increasing commitment to change of those hospitals which took part in the stress-management programme. However, stress, satisfaction, and commitment are all highly related, and the studies still demonstrated at an organizational level the very clear links between stress and clinical performance that we have reviewed in individuals. They also show the very real changes which can come about by the introduction of organizationally-based stress management programmes.

Conclusions

This chapter has reviewed the literature on the various links that might exist in the pathway from stressors such as a lack of sleep, through to performance outcomes such as accidents or incivility with patients. A summary of these pathways is shown in Fig. 9.1. From this we can see that sustained work and/or reduced sleep contribute both directly and indirectly, via the individual, to performance deficits, which can include cognitive and physical difficulties leading to accidents and poor relationships with colleagues and patients.

The stressors themselves include job-related events such as the death of patients or difficult colleagues which may themselves be adversely affected by tiredness. There is mounting evidence that these stressors do not act universally but depend on such things as the individual's personality, experience (both early family experience as well as job-related experience), and any recent life events. There is even some evidence that early life events, such as the death of a mother when the individual is still young, become serious issues for young doctors in terms of depression, although they might

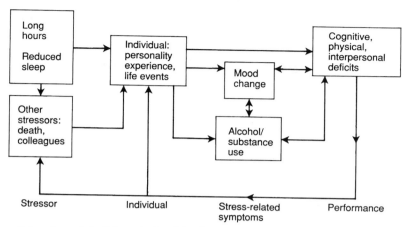

Figure 9.1   A model of the relationships between stress and performance.

not have shown effects earlier when they were students, perhaps because of the closer sense of responsibility for patients who die (Firth-Cozens, 1992a). As described, these factors may create mood changes, especially depression, either with or without the addition of alcohol, which will go on to have recognized cognitive and interpersonal deficits in terms of decision-making, concentration, and poor relationships. The cycle doesn't, however, stop there. Making mistakes or behaving badly to patients may stay with the doctor, sometimes for years, and this must be represented both by the memories of the mistakes becoming stressors in themselves, and by the effects upon the confidence and experience of the individual. Most of the literature reviewed has concerned junior doctors, though there are suggestions that depression and especially alcoholism is by no means restricted to those grades.

The research makes it clear that stress, in the widest sense of the word, is an important factor to be taken into account when looking at the performance of doctors and of the organizations who employ them. It is clear that aspects of performance that are affected by stress include both general occupational indices as well as those of mistakes and other less dramatic quality outcomes. However, there is now considerable evidence that stress and stress-related problems such as depression, anxiety, and alcohol abuse, can be treated by specialist job-related counselling and psychotherapy and this both reduces symptoms and changes job attitudes in a positive direction (Firth and Shapiro, 1986; Firth-Cozens and Hardy, 1992; Firth-Cozens, 1992b). As a result, a number of organizations have set up in-house counselling services, or provided help outside the organization. Studies such as Jones *et al.* (1988) described earlier, show how a stressed organization can as a whole reduce its 'symptoms'.

With so much at risk, it seems an essential aspect of the quality of care we offer to our patients that we should implement a programme for stress reduction at a number of levels:

1. committed training for undergraduates in ways to cope with stressors which are inevitable;

2. general stress-reduction programmes provided across the organization;

3. access to confidential counselling and psychotherapy services;

4. an acknowledgement of the evidence that sustained work and lack of sleep do affect the quality of patient care, as well as the individuals who provide it, and an introduction of more acceptable rota systems to enable adequate rest.

References

Alluisi, E. A. and Morgan, B. B. (1982). Temporal factors in human performance and productivity. In *Human performance and productivity. Volume 3: Stress and performance effectiveness* (ed. E. A. Alluisi and E. A. Fleishman), pp. 165–247. Lawrence Erlbaum Associates, Hillsdale, New Jersey.

Angus, R. G. and Heslegrave, R. J. (1985). Effects of sleep loss on sustained cognitive performance during a command and control simulation. *Behaviour Research Methods, Instruments and Computers* 17, 55–67.

Babkoff, H., Genser, S. G., Sing, H. C., and Thorne, D. R. (1985). The effects of progressive sleep loss on a lexical decision task: response lapses and response accuracy. *Behaviour Research Methods, Instruments and Computers* 17, 614–622.

Banks, M. H., Clegg, C, Jackson, P, Kemp, N. J., Stafford, E. M., and Wall, T. D. (1980). The use of the General Health Questionnaire as an indicator of mental health in occupational settings. *Journal of Occupational Psychology* 53, 187–94.

Bateman, T. S. and Organ, D. W. (1983). Job satisfaction and the good soldier: the relationship between affect and employee "citizenship". *Academy of Management Journal* 26, 587–95.

Bebbington, P. E., Hurry, J, Tennant, C, Sturt, E, and Wing, J. K. (1981). Epidemiology of mental disorders in Camberwell. *Psychological Medicine* 11, 561–80.

Brazier, W., Spurgeon, A., Hockey, R., Wiethoff, M., and Harrington, J. M. *The effects of fatigue on the health and performance of junior doctors* (unpublished).

Brooke, D., Edwards, G., and Taylor, C. (1991). Addiction as an occupational hazard: 144 doctors with drug and alcohol problems. *British Journal of Addiction* 86, 1011–16.

Carrere, S., Evans, G. W., Palsane, M. N., and Rivas, M. (1991). Job strain and occupational stress among urban transit operators. *Journal of Occupational Psychology* 64, 305–16.

Clark, D. C., Salazar-Grueso, E, Grabler, P., and Fawcett, J. (1984). Predictors of depression during the first 6 months of internship. *American Journal of Psychiatry* **141**, 1095–98.

Cohen, S. (1980). After effects of stress on human performance and social behaviour: a review of research and theory. *Psychological Bulletin* **88**, 82–108.

Crane, M. (1987, September 9). How lawyers pick the doctors they sue. *Medical Economics,* p.56.

Deary, I. J. and Tait, R. (1987). Effects of sleep disruption on cognitive performance and mood in medical house officers. *British Medical Journal* 1513–16.

Depue, R. A. and Monroe, S. M. (1986). Conceptualization and measurement of human disorder in life stress research. *Psychological Bulletin* **99**, 36–51.

Derogatis, L. R., Lipman, R. S., and Covi, M. D. (1973). SCL-90: an outpatient psychiatric rating scale — preliminary report. *Psychopharmacology Bulletin* **9**, 13–20.

Firth, J. and Shapiro, D. (1986). An evaluation of psychotherapy for job-related distress. *Journal of Occupational Psychology* **59**, 111–19.

Firth-Cozens, J. (1987a). The stress of medical training. In *Stress in health professional* (eds. R. L. Pyne and J. Firth-Cozens). Wiley, Chichester.

Firth-Cozens, J. (1987b). Emotional distress in junior house officers. *British Medical Journal* **295**, 533–36.

Firth-Cozens, J. (1990). Sources or stress in women junior house officers. *British Medical Journal* **301**, 89–91.

Firth-Cozens, J. (1992a). The role of early experiences in the perception of organizational stress: fusing clinical and organizational perspectives. *Journal of Occupational and Organizational Psychology,* **65**, 61–75.

Firth-Cozens, J. (1992b). Why me? A case study of the process of perceived occupational stress. *Human Relations* **45**, 131–41.

Firth-Cozens, J. and Hardy, G. (1992). Occupational stress, clinical treatment and changes in job perceptions. *Journal of Occupational and Organizational Psychology* **65**, 81–8.

Firth-Cozens, J. and Morrison, M. (1989). Sources of stress and ways of coping in junior house officers. *Stress Medicine* **5**, 121–26.

Ford, G. V. (1983). Emotional distress in internship and residency: a questionnaire study. *Psychiatric Medicine* **1**, 143–50.

Ford, C. V. and Wentz, D. K. (1986). Internship: what is stressful? *Southern Medical Journal* **79**, 595–99.

Friedman, R. C., Bigger, J. T., and Kornfeld, D. S. (1971). The intern and sleep loss. *New England Journal of Medical Education* **285**, 201–203.

Girard, D. E. and Hickam, D. H. (1991). Predictors of clinical performance among internal medicine residents. *Journal of General Internal Medicine* **6**, 150–54.

Glatt, M. M. (1976). Alcoholism: an occupational hazard for doctors. *Journal of Alcoholism* **11**, 85–91.

Goodwin, D. W. (1985). Alcoholism and alcoholic psychoses. In *Comprehensive textbook of psychiatry,* 4th edn. (ed. H. I. Kaplan and B. J. Sadock). Williams and Wilkins, Baltimore, Maryland.

Haslam, D. R. (1985). Sleep deprivation and naps. *Behaviour Research Methods, Instruments and Computers* **17**, 46–54.

Heath, R. G., Leach, B. E., Byers, L. W., Wartens, S., and Feighley, C. (1958). Pharmacological and biological psychotherapy. *American Journal of Psychiatry* **114**, 683–91.

Hsu, K. and Marshall, V. (1987). Prevalence of depression and distress in a large sample of Canadian residents, interns and fellows. *American Journal of Psychiatry* **144**, 1561–66.

Hurwitz, T. A., Beiser, M., Nichol, H., Patrick, L., and Kozak, J. (1987). Impaired interns and residents. *Canadian Journal of Psychiatry* **32**, 165–69.

Jex, S. M., Baldwin, J. R., Dewitt, C., Hughes, P., Storr, C., and Conrad, S. (1991). Behavioral consequences of job-related stress among resident physicians: the mediating role of psychological strain. *Psychological Reports* **69**, 339–49.

Jones, G. (1987). Stress in psychiatric nursing. In *Stress in health professionals* (eds. R. L. Payne and J. Firth-Cozens). Wiley, Chichester.

Jones, J. W., Barge, B. N., Steffy, B. D., Fay, L. M., Kunz, L. K., and Wuebker, L. J. (1988). Stress and medical malpractice: Organizational risk assessment and intervention. *Journal of Applied Psychology* **4**, 727–35.

Kinnunen, U. and Leskinen, E. (1989). Teacher stress during a school year: covariance and mean structure analyses. *Journal of Occupational Psychology* **62**, 111–22.

Kjelberg, A. (1977). Sleep deprivation and some aspects of performance, II lapses and other attentional effects. *Waking and Sleeping* **1**, 145–48.

Kreuger, G. P. (1989). Sustained work, fatigue, sleep loss and performance: a review of the issues. *Work and Stress* **3**, 129–41.

Lazarus, R. S., Deese, J., and Osler, J. F. (1952). The effects of psychological stress upon performance. *Psychological Bulletin* **49**, 293–316.

McAuliffe, W. E., Rohman, M., Breer, P., Wyshak, G., Santangelo, S., and Magnuson, E. (1991). Alcohol use and abuse in random samples of physicians and medical students. *American Journal of Public Health* **81**, 177–81.

McCue, J. (1985). The distress of internship: causes and prevention. *New England Journal of Medicine* **312**, 449–52.

Mackie, R. R. (1977). *Vigilance: theory, operational performance and physiological correlates*. Plenum Press, New York.

McManus, I. C., Lockwood, D. N. J., and Cruickshank, J. K. (1977). The pre-registration year. Chaos by consensus. *The Lancet* **i**, 413–17.

Mandell, H. and Spiro, H. (1987). *When doctors get sick*. Plenum Press, New York.

Margison, F. R. (1987). Stress in psychiatrists. In *Stress in health professionals* (eds. R. L. Payne and J. Firth-Cozens). Wiley, Chichester.

Merry, V. and Brown, G. I. (1987). *The neurotic behaviour of organisations*. Gardner Press, New York.

Mizrahi, T. (1984). Managing medical mistakes: ideology, insularity and accountability among internists-in-training. *Social Science and Medicine* **19**, 135–46.

Motowidlo, S. J. (1984). Does job satisfaction lead to consideration and personal sensitivity? *Academy of Management Journal* **27**, 910–15.

Motowidlo, S. J., Packard, J. S., and Manning, M. R. (1986). Occupational stress: its cause and consequences for job performance. *Journal of Applied Psychology* **71**, 618–29.

Mumford, E. (1983). Stress in the medical career. *Journal of Medical Education* **58**, 436–37.

Murray, R. M. (1976a). Alcoholism amongst male doctors in Scotland. *The Lancet* **2**, 729–32.

Murray, R. M. (1976b). Characteristics and prognosis of alcoholic doctors. *British Medical Journal* **2**, 1537–39.

Ormel, J. (1983). Neuroticism and well-being inventories: measuring traits or states? *Psychological Medicine* **13**, 165–76.

Payne, R. L. and Firth-Cozens, J. (1987). *Stress in health professionals*. Wiley, Chichester.

Pitts, F. N., Winokur, G., and Stewart, M. A. (1961). Psychiatric syndromes, anxiety symptoms and response to stress in medical students. *American Journal of Psychiatry* **118**, 333–40.

Poulton, E, Hunt, G., Carpenter, A., and Edwards, R. (1978). The performance of junior hospital doctors following reduced sleep and long hours of work. *Ergonomics* **21**, 279–95.

Reuben, D. B. (1985). Depressive symptoms in medical house officers. Effects of level of training and work rotation. *Archives of Internal Medicine* **145**, 286–88.

Rose, D. H. and Rosow, I. (1973). Physicians who kill themselves. *Archives of General Psychiatry* **29**, 800–805.

Rucinski, J. and Cybulska, E. (1985). Mentally ill doctors. *British Journal of Hospital Medicine* **33**, 90–94.

Scheibers (1987). Stress in physicians. In *Stress in health professionals* (eds. R. L. Payne and J. Firth-Cozens). Wiley, Chichester.

Schneider, B. and Bowman, D. E. (1985). Employee and customer perceptions of service in banks: replication and extension. *Journal of Applied Psychology* **70**, 423–33.

Sclare, B. (1979). Alcoholism in doctors. *British Journal of Alcohol and Alcoholism* **14**, 181–96.

Small, G. W. (1981). House officer stress syndrome. *Psychosomatics* **22**, 869–79.

Steppacher, R. C. and Mausner, J. S. (1974). Suicide in male and female physicians. *Journal of the American Medical Association* **228**, 323–28.

Uliana, R. L., Hubbell, F. A., Wyle, R. A., and Gordon, G. H. (1984). Mood changes during internship. *Journal of Medical Education* **59**, 118–23.

Vaillant, G. E., Brighton, J. R., and McArthur, C. (1971). Physicians use of mood-altering drugs: a 20 year follow-up report. *New England Journal of Medicine* **282**, 365–70.

Valko, R. J. and Clayton, P. J. (1975). Depression in the internship. *Diseases of Nervous System* **36**, 26–29.

Warm, J. S. (1984). *Sustained attention in human performance*. Wiley, New York.

Webb, W. B. and Levy, C. M. (1984). Effects of spaced and repeated total sleep deprivation. *Ergonomics* **27**, 45–58.

West, M., Jones, A., and Savage, Y. (1988). Stress in health visiting: a quantitative assessment. *Health Visitor* **81**, 269–71.

Williams, H. L., Lubin, A., and Goodnow, J. J. (1959). Impaired performance with acute sleep loss. *Psychological Monographs* **73**, 1–26.

10

Recovering from a medical accident: the consequences for patients and their families

CHARLES VINCENT and IAN H. ROBERTSON

A medical accident, like other similar traumas, may affect every aspect of a person's life. It can lead to pain and disability, which in turn may affect employment prospects and financial security. There may also be psychological distress and changes to relationships with family and friends. Research on the effects of accidents of other kinds suggests that the psychological effects of an injury may be both sustained and severe (Malt, 1988; Landsman et al., 1990). In cases where the victim of an accident is not to blame, such as a medical accident, the psychological trauma is likely to be heightened (Brewin, 1984). Some adverse events may produce minimal physical disability, but have marked psychological consequences. Awareness during anaesthesia is one example of this, but many less traumatic experiences may also have profound consequences.

There is only one study (Vincent et al., 1993), to our knowledge, that explores the wider effects of medical accidents on the patients involved. We shall therefore also review literature on the effects of other kinds of accident, but consider what difference it makes to a patient that they were injured in hospital rather than say, in a road accident. The situation is unusual in that the patient may be injured by the very people who were trying to help. Their reactions may be especially powerful and have a considerable impact on their relationships with health professionals. Staff involved may also have difficulty in knowing how to react in such a situation.

The nature of adverse events

The Harvard Medical Practice Study (Hiatt et al., 1989) described a number of different types of adverse events in medicine, defined as those leading to measurable disability. The most common were drug complications, wound infections, complications during surgery, and delayed diagnosis. The degree and duration of disability, over and above pre-existing illness, was rated by the reviewing physicians and later in a telephone interview. Pain, psychological trauma, and the effects on family and social life were not directly assessed. The great majority of adverse events did not results in serious disability. More than half the patients had minimal impairment, recovering completely in a month or less. Seventy per cent recovered completely in less than six months. However, 13.6 per cent of patients suffering adverse events died and 6.5 per cent sustained a permanent disability. If we apply these findings to the approximately eight million admissions per year in British hospitals (DHSS, 1991), this suggests that about 40 000 deaths and 19 000 cases of permanent disability would be due wholly or partly to medical intervention.

Diagnostic mishaps were associated with higher disability than other forms of adverse event. Ectopic pregnancy was the most frequently missed diagnosis, with delayed diagnosis of appendicitis and cancer being other common errors. Non-technical complications of surgery (such as pulmonary embolism, heart attack, or stroke) were more likely to result in serious disability than purely technical complications such as bleeding or wound problems. Drug complications led to a relatively low proportion of serious disability. Finally we should note that a small proportion of adverse events result in brain damage to the patient, the most obvious example, and one that attracts most attention, is that of a baby becoming hypoxic during labour.

The psychological consequences of accidents and trauma

Most clinical studies of people involved in road traffic accidents (with or without ensuing litigation) have been concerned primarily with the assessment and treatment of the physical injuries. These studies, which usually involve more severely injured patients, suggest that 20–30 per cent of patients suffer long-term psychological impairment; anxiety disorders, depression, and organic brain syndromes are the most common. However, psychological consequences have not been the major focus and the assessment of psychological damage has been generally been rudimentary (Malt, 1988).

A small number of studies have focused specifically on the psychological consequences of accidents. Shepherd *et al.* (1990) assessed patients with fractures of the jaw sustained in assaults or accidents. At three months 19 per cent of accident victims were still considered to be at risk of psychiatric disorder and 17 per cent remained depressed. There were, however, few details of the extent of the injuries or the nature of the distress these patients experienced. Jones and Riley (1987) studied 327 subjects involved in civil accident litigation. Only a minority had problems that were sufficiently severe to warrant a psychiatric diagnosis, but most had symptoms of depression, anxiety, irritability, and sleep disturbance. Almost half were considered to be pre-occupied with the injury they had suffered.

Malt (1988) studied a random sample of patients presenting to a hospital accident department with accidental injuries. These injuries were of widely varying severity; 24 per cent were considered to be serious, but only 9 per cent of patients stayed in hospital more than two weeks. Six months after the accident 17 per cent of patients were judged to have a psychiatric disorder caused by the accident, reducing to 9 per cent at two years. Depression and anxiety disorders were the most common problems, but only one case of post-traumatic stress disorder was recorded. Malt comments that the levels of psychiatric disorder are very much lower than those seen after major disasters, and that post-traumatic stress disorder is very much rarer.

Landsman *et al.* (1990) studied 137 patients with more severe injuries, from road accidents, falls, pedestrian accidents, stabbing, and gunshot wounds. Questionnaires were administered sixteen months after the accident, but only 10 per cent of those approached agreed to take part. Accident victims recorded a high frequency of intrusive images and memories of the event and higher levels of psychiatric symptomatology than those for the general population. The extremely low response rate means that the study cannot provide a reliable assessment of the incidence of disorder; the interest lies in the analysis of the factors that predicted high distress. Time since the accident, and the initial severity of injury, did not predict long-term distress. The resultant disability, which is probably a more important predictor of long-term adjustment, was not assessed. Long-term distress was associated with problems at work and a poorer family environment, though the direction of causation was not clear. An unsupportive family may delay recovery and exacerbate distress; however a very distressed person can also affect the atmosphere in a family.

In summary, people involved in accidents frequently suffer from a variety of psychological problems, particularly depression, anxiety, and disturbing and intrusive memories of the accident. Only 10–20 per cent of people are considered to have a psychiatric disorder caused by the accident, but many of these continue to be distressed for months or years afterwards.

Some medical accidents, for example those occurring during surgery, are relatively sudden and may involve injuries similar to those from other accidents. With others, such as delays in diagnosis, the injury may be much less clearly defined. We need to be cautious about extrapolating from accidental injuries to medical accidents. However, there is sufficient evidence from the accident literature to suggest that medical accidents probably have important psychological effects on those involved, in addition to any physical injury or pathology.

The consequences of medical accidents

To our knowledge only one study has focused on the effects of medical accidents. Vincent *et al.* (1993) gave questionnaires to 101 patients who had contacted Action for Victims of Medical Accidents (AVMA) with a question or complaint relating to their surgical treatment. These patients had all been sufficiently dissatisfied or distressed by their treatment to have considered litigation, and over half had decided to proceed. The study examined: (a) the effects it had on their life; (b) explanations given pre- and post-operatively; (c) whether they had received an admission or responsibility or apology; (d) whether they felt the incident was preventable and, if so, who they blamed; and (e) whether they had considered litigation and their reasons for proceeding or not proceeding. Blame, avoidability and adequacy of pre- and post-operative explanations were also rated. Standardized questionnaires consisted of: (a) the General Health Questionnaire (GHQ-28) (Goldberg and Hillier, 1979) to give an indication of the presence or absence of psychiatric disorder; (b) Impact of Events Scale (IES), which assesses the effects of traumatic events, such as accidents and serious life events (Horowitz, 1979); (c) McGill Pain Questionnaire (Short form) (Melzack, 1987); and (d) Psychological Adjustment to Illness Scale (PAIS) (Derogatis, 1986). The PAIS is a wide-ranging measure that assesses several aspects of a person's adjustment to their illness or disability. The categories concern attitudes to health care, effect on studies or work, effect on home life, sexual relationships, family relationships, social life, and psychological distress.

The medical records of these patients were not available, but they provided descriptions of the accident (as they saw it) which were reviewed by a consultant surgeon. Injuries to an organ or nerve damage accounted for over a third of cases. In another third of cases the patient complained of increased pain, but the exact nature of the damage was unclear. The remainder included perforations, wound infections, and serious disability. In a third of cases the surgeon was unable to assess the extent of damage and standard of treatment. In the remainder he found that three quarters of the patients had injuries caused by treatment and in half the treatment was negligent.

The overall effect on the patient's lives, as judged by them, was considerable. For 30 per cent of patients the main effect was physical (primarily increased pain or decreased mobility), and for 16 per cent the trauma was primarily psychological. The remainder cited a combination of factors: physical, psychological, social, and financial. On average, medical accident patients had, after 16 months, levels of pain comparable to those of unmedicated patients recovering from surgery. Seventy four per cent of patients rated the overall effect on their lives as severe or very severe, and 35 per cent experienced severe financial difficulties.

Scores on the General Health Questionnaire suggested that 77 per cent of these patients could be suffering from a psychiatric disorder. This does not mean that these patients had a pre-existing psychiatric disorder, simply that their distress was sufficiently high to warrant a probable psychiatric diagnosis. Patients also suffered from frequent distressing memories, as shown

Table 10.1 Psychological impact of medical accidents: mean scores (S.D.) on the Impact of Events scale

Impact of events scale	Medical accidents	Serious life events	Accidents
Intrusion	21.3 (9.2)	21.4	5.48
Avoidance	18.5 (10.4)	18.2	9.26
Total	39.8 (17.2)	39.5	14.74

by scores on the Impact of Events Scale (Table 10.1). Scores from other accidentally injured people (mostly road accidents) (Malt, 1988) and for people suffering serious life events (bereavements and assaults) (Horowitz *et al.*, 1979) are also shown to indicate the severity of the trauma suffered by the medical accident patients.

Scores on the individual subscales of the PAIS are shown in Table 10.2. PAIS scores for patients with gynaecologic cancer (Cain *et al.*, 1986) and for patients with coronary heart disease both pre-operatively and twelve months after bypass surgery (Langeluddecke *et al.*, 1989) are also presented for comparison.

Psychosocial adjustment in this group of patients was extremely poor more than a year after the surgery. Many of them would have been suffering from pre-existing illness, and this has no doubt contributed to their distress. However, their overall adjustment is markedly worse than patients with severe and even life-threatening illnesses, particularly with regard to their social activities, work lives, and attitude to health care. It could be argued that these patients might be exaggerating their symptoms in the hope of obtaining compensation. However, they were told that their replies would be confidential and would have no bearing whatever on any litigation in which they were involved. There was also no difference in the levels of distress between patients who had decided not to proceed with litigation, who would presumably have had no reason to exaggerate, and those who were going ahead. However, the authors stress that the study has a number of limitations. Firstly, these patients are a highly selected group; it is likely that they are the more seriously affected. Secondly, the sample is small and the injuries and traumas experienced are of many different kinds. Thirdly, there was not sufficient medical information to be able to judge the severity of the physical injury, and the extent to which medical treatment fell below acceptable standards.

Table 10.2 Psychological adjustment: mean scores (S.D.) on the individual subscales of the Psychological Adjustment to Illness scale

	Medical accidents	Coronary heart disease (pre-operatively)	Coronary heart disease (at 1 year)	Cancer
Health care	12.2 (4.4)	11.5	5.1	7.6
Vocational	7.4 (4.8)	11.3	2.0	6.8
Domestic	8.3 (4.4)	10.9	3.9	5.3
Sexual	6.9 (5.0)	7.6	6.0	10.1
Family	2.7 (2.8)	1.9	1.6	1.4
Social	9.1 (5.1)	8.8	5.8	4.4
Distress	10.6 (5.0)	6.4	4.0	6.8
Total	57.2 (21.6)	58.5	28.4	42.4

There were indications that the physical effects of the injury only partly accounted for the high distress. Distress was still high and adjustment poor even when injuries were less severe, although some patients may not have communicated the full extent of their injuries. Secondly, the assessment of the patients' views of the explanations they received after surgery suggests that the way the trauma was handled afterwards may be an important factor.

Brain damage after medical intervention

The brain is as vulnerable as it is precious. Starved of oxygen, neurones will die and will not regenerate. Even subtle effects of such damage can have a severe impact on a person's life. The effects of major damage to this organ are seldom less than catastrophic. It is unfortunately the case that many medical procedures can, directly or indirectly, place the brain at risk for a reduction in either the quantity or quality of its blood supply, thus raising the spectre of brain damage. Examples include very low or very high blood pressure, severe dehydration, inadequate oxygen during anaesthesia, poor blood filtration in bypass surgery, peri-operative strokes, perinatal difficulties, and many others.

Coronary artery bypass graft surgery is one procedure in which the brain appears to be at risk. In a study carried out in Newcastle (Shaw *et al.,* 1986a,b), 259 people who had undergone this operation were assessed 7 days and then 6 months post-operation. At 7 days, the majority (79 per cent) showed significant cognitive impairment, of which 24 per cent were considered to be 'moderate or severe'. At 6 months, 57 per cent of the sample showed some evidence of cognitive impairment, though only 6.6 per cent were found to have 'moderate or severe' cognitive deficits.

Memory and attention problems predominated among the deficits found: these functions are also most commonly impaired following closed head injury, and memory/attention difficulties often cause marked problems in everyday life activities, including work. Of the 147 people who showed impairment at 6 months, only 40 complained specifically of cognitive problems, and these tended to be of forgetfulness, mental slowness, and reduced ability to concentrate, which again are the classic complaints of closed head injury sufferers. The fact that the majority of people who showed test evidence of cognitive deficits did not report cognitive difficulties in everyday life does not of course mean that such difficulties did not exist, only that they were unacknowledged by the individuals concerned.

Even taking the strict criterion of a combination of objective test evidence of impairment together with self-reported cognitive difficulties, these studies suggest that around 15 per cent of coronary artery bypass graft patients in this sample could be considered to be significantly mentally impaired following their operation.

The confusion and distress described earlier in this chapter as arising from disability and pain caused by medical accidents which are not explained or accounted for may be multiplied many times where the very disability experienced by the patient is denied. This can be the case where standard neurological examination fails to reveal neurological abnormality, yet the patient complains of poor memory and concentration, for instance, or where the family complain of personality change.

The possibility of undetected cognitive deficits is increased by the fact that such deficits are usually accompanied by a significant degree of psychological distress, including depression. Complaints of poor memory and concentration are therefore often attributed to emotional factors rather than to damaged brain tissue. While memory, concentration, and other cognitive deficits are indeed at times attributable to severe emotional distress, it is also the case that routine medical and neurological screening is simply not oriented towards detecting all but the most gross deficits in cognitive functioning. Neuro-psychological assessment, on the other hand, based on psychometrically sophisticated procedures, can often detect changes in cognitive functioning which otherwise would be missed. The coronary artery bypass graft studies mentioned above are an example of this. Unfortunately, however, neuropsychologists are rarely available in most medical settings.

To give an example of the kind of case that often goes undetected, a sixteen-year-old boy seen by one of the authors was referred following an operation to his ankle which resulted in severe pain for several months afterwards The boy's complaint was about the emotional distress caused by the pain during several months before this was acknowledged by the responsible surgeons, and a corrective operation carried out.

However, the boy's father mentioned in passing that his son's personality had changed since the accident, and that he had become somewhat lazy, disorganized and short-tempered, though he was still holding down a place in an engineering apprentice training scheme, could converse appropriately, and showed no obvious problems on interview. He also mentioned that the boy had taken much longer to wake up from the anaesthetic after this operation than after previous identical operations with the same anaesthetic procedure. Neuropsychological assessment found clear evidence of a small but highly significant decline in intellectual abilities, as well as marked difficulties in new learning and in concentration.

These problems were compatible with the reports of personality change, and closer investigation of the anaesthetic procedure by an independent anaesthetist revealed a number of abnormalities which, among other things, very likely resulted in a period of prolonged low blood pressure, as well as an extended period of dehydration, both of which probably contributed to a moderate degree of brain damage which nevertheless had far-reaching and long-lasting effects on the boy's life.

The long-term consequences of cognitive impairment after medical procedures are hard to assess as there is little research in this area. Extrapolating from the studies of Shaw *et al.* (1986a,b), as well as from the literature on closed head injury, a picture of problems with memory, concentration, personality, and interpersonal relationships appears (e.g. McKinlay *et al.*, 1981), even after quite moderate brain injuries which may be insufficiently severe to show damage with closed head injury suggests a high degree of emotional distress, social isolation, poor employment prospects, and indeed a tendency for problems to become worse rather than better over time (e.g. Thomsen, 1984), though few medical accidents will produce symptoms of this severity. The inclusion of brief neuropsychological assessments for those cases where insult to the brain may have been a possibility during medical procedures is essential for future research in this area.

Brain damage can of course also occur around the time of birth of a child, again for a variety of reasons which will not be considered here. It would be facile to labour the degree and disability which perinatal brain damage causes both child and family over the course of their lives, as it would be to emphasize the different effects of brain damage on the developing child compared to those on the mature adult. One study has suggested that perinatal factors may play a part in 30 per cent of cases of cerebral palsy (Hagberg *et al.*, 1989). To attempt to review the area would be a travesty of the enormous literature on developmental problems in children brain-injured at birth, and the reader is referred to Aicardi (1992) for a full review.

When a patient dies

The death of a patient primarily affects their immediate family, though the repercussions may be widespread. To our knowledge there are no studies which specifically examine the course of bereavement in relatives of patients whose deaths were in part a consequence of incorrect medical treatment. However, we can consider studies of bereavement after accidents of other kinds or other unexpected deaths. The effect on a patient who discovers that he is dying and that his treatment has been incorrect can only be guessed at.

Early studies on bereavement tended to view it as a relatively transient reaction, lasting a few weeks or months. Although symptoms do decline with time, recent and more thorough studies have found that an uncomplicated bereavement usually lasts at least a year and that marked symptoms are often present up to four years after the death (Lehman *et al.*, 1987). For instance Vachon *et al.* (1982) found that 38 per cent of bereaved widows were experiencing high levels of distress after one year. Parkes and Weiss (1983), in a longitudinal study, found that 40 per cent of widows and widowers showed moderate or severe anxiety over two years after the loss. Parkes (1988) has emphasized that any bereavement involves multiple losses: the widow or widower loses the companionship, a confidant, their sexual relationship, and may experience a loss of identity. Many bereaved people describe the loss in almost physical terms — as having part of them torn away. There may also be financial difficulties and a variety of practical problems.

Available evidence suggests that bereavement is particularly likely to have a long-lasting impact if the loss is untimely or unexpected. Younger widows and widowers generally experience more profound distress than those who lose their spouse at a later age (e.g. Vachon *et al.*, 1982). Most evidence suggests that the effects of bereavement are more severe where the bereaved has had little forewarning about the loss. For example, Parkes and Weiss (1983) found that two years after the loss only one of 18 (6 per cent) spouses with brief forewarning was rated as 'doing well', compared with 63 per cent of those with longer forewarning. When Lundin followed up first degree relatives eight years after bereavement he found that the unexpectedly bereaved were still more tearful, self-reproachful, and numb and that they missed the dead person and mourned for them more than other people (Lundin, 1984).

There are indications that a bereavement that follows a sudden, accidental death may be exceptionally severe. Lehman *et al.* (1987) studied people four to seven years after they had lost a spouse or child in an accident. They found significant differences between bereaved spouses and controls in levels of depression and other psychiatric symptoms, social functioning, psychological well-being, reactions to good events, and future worries and concerns. The findings for parents of children who had died were similar, though their long-term problems did not appear to be as severe. Thoughts and feelings about the deceased person continued to occupy the thoughts and conversations of bereaved spouses and parents. Moreover, a large percentage continued to ruminate about the accident and what could have been done to prevent it, and they appeared unable to accept, resolve, or find any meaning in the loss. The authors comment that 'the data provide little support for the traditional notions of recovery from the sudden, unexpected loss of a spouse or child'.

The Harvard study found that 13 per cent of adverse events involved the death of a patient, but they made it clear that the majority of these patients were already seriously ill. Treatment may have hastened or precipitated the death, but it was rarely the whole cause. Nevertheless, if the death was accidental, and especially if it was sudden, we might expect the course of the subsequent bereavement to be difficult and, if the death was avoidable, exceptionally severe.

Factors affecting recovery

The impact of the accident and the speed and extent of recovery will depend on a considerable number of different factors. Clearly the nature and extent of the injury, the level of pain, and the degree of subsequent disability are crucial. The personality of the patient involved, the history of previous trauma and loss in their life, their financial security, and employment prospects may also influence subsequent adjustment. In this section we shall focus on those factors which seem to us to be especially relevant to recovery from medical accidents. We believe that they all have important clinical and sometimes legal implications, and need to be considered in the care of patients injured during treatment.

The nature of the accident. Traumatic and life-threatening events may produce a variety of symptoms, over and above any physical injury. Anxiety, intrusive memories, emotional numbing, and flashbacks are all common sequelae of such events and are important components of post-traumatic stress disorder. Some medical accidents may be sudden, dangerous, and frightening, but resolve relatively quickly. Awareness under anaesthesia is an example of such an event. In these cases the nature of the actual event may be the main predictor of subsequent distress. Rachman (1980) views such symptoms as evidence of failures of the 'emotional processing' that follows any stressful event. Sudden, intense, dangerous, or uncontrollable events are particularly likely to be lead to such problems, especially if accompanied by illness, fatigue, or mood disturbances. Vincent *et al.* (1993) found that patients report frequent disturbing intrusive memories of the events surrounding the accident.

Most accidents, and probably also medical accidents, do not produce post-traumatic stress disorder in its pure form. In most cases the long-term conse-quences of the event, in terms of pain, disability, and effect on family and other relationships will be much more important. Depression is much more likely to follow events that carry 'long-term threat', than briefer traumas no matter how severe (Brown and Harris, 1978). Whether people actually become depressed and to what degree will depend on the severity of their pain and disability, the support they have from family, friends and health professionals, and a variety of other factors.

Explanations after medical accidents. Many studies have shown that patients are generally dissatisfied with the information that they are given in the ordinary course of treatment (Ley, 1989). The lack of a clear and convincing explanation may be especially distressing after something has gone wrong. Action for Victims of Medical Accidents (AVMA) has suggested that many patients turn to litigation primarily because they failed to obtain a clear explanation of what has happened (Simanowitz, 1985).

Vincent *et al.* (1993), in the study described above, found that 20 per cent of patients injured during treatment waited longer than six months for an explanation. In only 21 per cent of cases was responsibility accepted for the incident, and in only 27 per cent was an apology offered. In 34 per cent of cases no one else was present, the patient being supported neither by other staff nor by relatives. Eighty one per cent were dissatisfied or very dissatisfied with the amount of information they were given, 66 per cent were dissatisfied with its clarity, 62 per cent with its accuracy (as they perceived it), and 62 per cent felt that it was given unsympathetically. Given the lack of clarity of the explanations it is particularly alarming that almost half the patients (44 per cent) reported that they had no opportunity to ask questions.

Poor explanations were associated with a higher frequency of disturbing memories over a year later, and with problems at work and socially. The association between distress and adequacy of explanation is important, but care must be taken in its interpretation. It may well be that the lack of a clear and sympathetic explanation actually increases these patients' distress. Alternatively, subsequent distress and anger might bias the patient's recollec-tion of the explanation they were given. The findings show at the very least that a clear explanation is seen as extremely important by patients consider-ing litigation and that they are extremely dissatisfied with the explanations they have been given. The lack of a clear explanation may also reduce their trust in doctors and health professionals. This can lead them to avoid having further treatment — which in most cases they very much need.

Support from family, friends, and health professionals. Social support refers to the perceived comfort and caring or practical help a person receives from other people. It may come from their spouse, family, friends, colleagues, doctors, or community organizations. A crucial determinant of the support people receive in times of crisis is the extent of their social network and the attitudes of those around them. The way they react to a crisis will also affect the willingness of those around to help.

Social support has been found to affect a person's response and adjustment to a wide range of stressful events. The support a person receives affects the amount of stress they suffer in their work, how they cope with a young baby, and how they cope with chronic illness (Sarafino, 1990). Social support was also found, in a nine-year prospective study, to affect mortality from cancer and heart disease (Berkman and Syme, 1979). The likelihood of a person becoming depressed after a serious life event is very much affected by the presence or absence of a close, confiding relationship (Brown and Harris, 1978). Landsman *et al.* (1990) found that the family environment was related to the adjustment of people to a traumatic injury.

These findings suggest that the reaction to a medical accident will be partly determined by the support patients have from those around them. An especially important source of support will be the doctors and other health professional who are involved in their treatment. Patients who have been injured during their treatment may need more time and support than other patients. Both patients and doctors seem to change their attitudes to each other after medical accidents. Vincent *et al.* (1993) found that three quarters of the patients in their study reported that their attitude to the medical profession had changed as a result of their experiences. Some had less confidence in doctors' competence, while others expressed their change in attitude in more personal terms, mostly indicating that they had less trust in doctors than previously. Over half of this group reported that the attitude of staff to them had changed after the incident. Of these, 43 per cent found that the staff had been more attentive and caring, but 54 per cent found the staff's attitude changed for the worse. Typical comments were that the staff were more withdrawn and distant, and gave them less information.

Blame. Studies of accident victims (Jones and Riley, 1987) have found that many express considerable bitterness about their injuries. They may become preoccupied with the fact that another person is to blame for their misfortune. Vincent *et al.* (1993) found that the majority of medical accident victims in their study blamed their doctors, and that degree of blame was associated with the frequency of disturbing memories. The relationship between blame and subsequent litigation is uncertain (Lloyd Bostock, 1980) but there is growing evidence that blaming another person affects subsequent adjustment.

The relationship between blaming another for a threatening event and subsequent adaptation has recently been explored by Tennen and Affleck (1990) in a comprehensive review. Of the 22 studies where participants viewed another person as the cause of their misfortune or accident, 17 found that blame was related to subsequent adaptation. For instance women who blamed doctors for a fetal or perinatal death were more likely to feel cheated and view the death as unfair (Graham *et al.*, 1987). Mothers who blamed others for their children's severe medical problems showed greater mood disturbance and care-taking problems (Tennen and Affleck, 1990).

Tennen and Affleck propose a model in which blame is likely if the outcome is severe and the event or accident is associated with the clear involvement of one person of high authority, especially if they are not well known or well liked. This suggests that an adverse outcome during medical treatment is an occasion where, rightly or wrongly, blame is highly likely to be attached to the doctor involved, especially if there is not a good rapport between doctor and patient. Tennen and Affleck suggest that blaming others interferes with adjustment in three main ways. Firstly, preoccupation with blaming the other person may interfere with efforts at coping with the trauma or loss, and distract the person from efforts to solve their problems. Secondly, blaming others may challenge cherished and deeply held views, which may in itself be threatening. For instance, people with serious illnesses may place a great deal of faith in the ability of their doctors to help them; to blame them for failing is to leave oneself feeling vulnerable and unprotected. Thirdly, blaming others may deprive one of the support of other people. To blame the doctor who is treating you, or worse—to blame health professionals generally, may deprive the patient of support and help at the very point when he or she needs it most.

Litigation. Although there is widespread concern about the level of litigation in many countries, it is a relatively infrequent response to a medical accident (see Chapter 2). However, where litigation or compensation are involved, they certainly may affect post-accident adjustment.

Firstly the actual legal process itself is expensive, protracted, and stressful for all concerned. The patient is faced with many additional interviews and medical tests. The opposing lawyers and experts may attempt to cast doubt on the patient's account, which may anger and distress them. The whole process functions as a continual reminder of the original incident and may interfere with the patient's attempts to return to a more normal life. As against that a successful action may, in addition to bringing financial benefits, make the patient feel that they are finally being believed which may come as a considerable relief.

More controversial is the suggestion that if litigation is underway symptoms may be prolonged, exaggerated, or sometimes invented in order to obtain compensation. If this is a conscious and deliberate deceit, it is referred to as malingering. More commonly, the suggestion is that the patient is unaware of doing this and is suffering from 'compensation neurosis'.

Dworkin *et al.* (1985) have reviewed the empirical work that relates to compensation, litigation, and chronic pain. Chronic pain patients who have litigation pending do not differ from those without litigation pending with respect to non-organic signs, their description of their pain, psychological distress, and various indicators of pain behaviour. There are some studies suggesting that chronic pain patients with litigation pending have a significantly poorer response to treatment, but an equal number show no difference. In their study Dworkin *et al.* (1985) found that compensation status and litigation were secondary to employment status as predictors of short-term treatment outcome, and only employment status predicted long-term outcome. They concluded that 'it would be valuable to redirect attention away from the deleterious effects of compensation neurosis and towards the role of activity and employment in the treatment and rehabilitation of chronic pain patients'. They also comment that there is little evidence that patients typically resume employment when litigation following industrial and automobile injuries is settled (Mendelson, 1984; Dworkin *et al.*, 1985).

There is as yet no research examining recovery from medical accidents. These studies suggest that compensation and litigation should probably not be considered as primary determinants of the recovery process, and that 'compensation neurosis' is comparatively rare.

The needs of patients and their families after medical accidents

When a patient has been injured during medical treatment they obviously may need further, remedial treatment. They may also need financial help or compensation, which is discussed elsewhere in this book. The studies discussed here indicate that the psychological and social needs of these patients also need to be taken seriously by those involved in their care. If these needs are addressed, the long-term outcome for patients will be better, and litigation may be less likely. We suspect that many of these patients are in any case not primarily seeking compensation. They want an explanation, an acknowledgement that something has gone wrong, and an assurance that some action will be taken to prevent other people suffering.

Firstly, we suggest that patients who consider that they have been injured during their treatment should in the first instance be believed. In many cases it may turn out that they had unrealistic expectations of their treatment, or had not fully understood the risks involved. In a few cases they may be malingering or hypochondriacal. However, if 4 per cent of admissions result in some kind of injury to the patient, as the Harvard study suggests, then a report of such an injury should at least be seen as credible. It should certainly not be automatically seen as evidence of personality problems, or of being 'difficult'. In our experience being believed is extremely important for accident victims and, conversely, not being believed is extremely frustrating and distressing — irrespective of whether there is litigation involved.

Secondly, patients involved in accidents need to understand what has gone wrong. It is possible that an inadequate explanation actually increases long-term distress and, where a patient has died, affects the course of a bereavement. The distress caused by an accident is likely to be less if it can be made intelligible: the fallibility of both human beings and human systems is well known to ordinary people, and many may accept their injury in this context. However, when patients think that information is being concealed from them, or that they are being dismissed as trouble makers, then it is much more difficult for them to make sense of their injury, and may hence be more prone to litigation. We suggest that when something has gone wrong a senior doctor needs to give a thorough and clear account of what exactly happened. The patient and their relatives need to have time to reflect on what was said and to be able to return and ask further questions. Similar considerations of course apply when doctors are breaking bad news of any kind.

Thirdly, a proportion of patients are likely to be sufficiently anxious or depressed to warrant formal psychological or psychiatric treatment. Sometimes this may be best dealt with by referral to the appropriate specialists, though the patient may well be wary of their problems being seen as 'psychological' or 'all in the mind'. We suggest that a specialist counselling service may be needed to meet the needs of some medical accident victims. This would be of benefit both to injured and traumatized patients and to the staff who care for them.

References

Aicardi, J. (1992). *Diseases of the nervous system in childhood.* MacKeith Press, London.
Berkman, L. F. and Syme, S. L. (1979). Social networks, host resistance and mortality: a nine year follow-up study of Alameda County residents. *American Journal of Epidemiology* **109**, 186–204.
Brewin, C. R. (1984). Attributions for industrial accidents: their relationship to rehabilitation outcome. *Journal of Social & Clinical Psychology* **2**(2), 156–64.
Brown, G. W. and Harris, T. (1978). *Social origins of depression.* Tavistock Publications, London.

Cain, E. N., Kohorn, E. I., Quinlan, D. M., Latimer, K., and Shwartz, P. E. (1986). Psychosocial benefits of a cancer support group. *Cancer* 57, 183–89.

Derogatis, L. R. (1986). The psychosocial adjustment to illness scale (PAIS). *Journal of Psychosomatic Research* 30(1), 77–91.

DHSS (1991). *Hospital inpatient enquiry summary tables.* HMSO, London.

Dworkin, R. H., Handlin, D. S., Richlin, D. N., Brand, L., and Vannucci, C. (1985). Unravelling the effects of compensation, litigation and employment on treatment response in chronic pain. *Pain* 23, 49–59.

Goldberg, D. P. and Hillier, V. F. (1979). A scaled version of the General Health Questionnaire. *Psychological Medicine* 9, 139–45.

Graham, M. A., Thompson, S. C., Estrada, M., and Yonekura, M. L. (1987). Factors affecting psychological adjustment to a fetal death. *American Journal of Obstetrics and Gynaecology* 157, 254–57.

Hagberg, B., Hagberg, G., Olow, J., and Von Wendt, L. (1989). The changing panorama of cerebral palsy in Sweden V: the birth year period 1979–1982. *Acta Paediatrica Scandinavia* 78, 283–90.

Hiatt, H. H., Barnes, B. A., and Brennan, T. A. (1989). A study of medical injury and malpractice. *New England Journal of Medicine* 321, 480–84.

Horowitz, M., Wilner, N., and Alvarez, W. (1979). Impact of event scale: a measure of subjective stress. *Psychosomatic Medicine* 41(3), 209–19.

Jones, I. H., and Riley, W. T. (1987). The post-accident syndrome: variations in the clinical picture. *Australian and New Zealand Journal of Psychiatry* 21, 560–67.

Landsman, I. S., Baum, C. G., and Arnkoff, D. B. (1990). The psychosocial consequences of traumatic injury. *Journal of Behaviourial Sciences* 13(6), 156–64.

Langeluddecke, P., Fulcher, G., Baird, D., Hughes, C., and Tennant, C. (1989). A prospective evaluation of the psychosocial effects of coronary artery bypass surgery. *Journal of Psychosomatic Research* 33(1), 37–45.

Lehman, D. R., Wortman, C. B., and Williams, A. F. (1987). Long-term effects of losing a spouse or child in a motor vehicle crash. *Journal of Personality and Social Psychology* 52(1), 218–31.

Ley, P. (1989). Improving patients' understanding, recall, satisfaction and compliance. In *Health psychology: processes and applications* (ed. A. K. Broome). Chapman and Hall, London.

Lloyd Bostock, S. M. A. (1980). Fault and liability for accidents: the accident victim's perspective. In *Compensation for illness and injury* (ed. C. Harris). Oxford University Press.

Lundin, T. (1984). Morbidity following sudden and unexpected bereavement. *British Journal of Psychiatry* 144, 84–88.

McKinlay, W. M., Brooks, D. N., Bond, M. R., Martinage, D., and Marshall, M. (1981). The short term outcome of severe blunt head injury as reported by relatives of the head injured person. *Journal of Neurology, Neurosurgery and Psychiatry* 44, 527–33.

Malt, U. (1988). The long-term psychiatric consequences of accidental injury. A longitudinal study of 107 adults. *British Journal of Psychiatry,* 153, 810–18.

Melzack, R. (1987). The short-form McGill Pain Questionnaire. *Pain* 30, 191–97.

Mendelson, G. (1984). Follow-up studies of personal injury litigants. *International Journal of Law and Psychiatry* 7, 179–88.

Parkes, C. M. (1988). *Bereavement: studies of grief in adult life* (2nd edn). Penguin, London.

Rachman, S. (1980). Emotional processing. *Behaviour Research and Therapy* **18**, 51–60.

Sarafino, E. P. (1990). *Health psychology*. Wiley, New York.

Shaw, P. J., Bates, D., Cartilidge, N. E. F., French, J. M., Heaviside, D., Julian, D. G., and Shaw, D. A. (1986a). Early intellectual dysfunction following coronary bypass surgery. *Quarterly Journal of Medicine (New Series)* **58**, 59–68.

Shaw, P. J., Bates, D., Cartilidge, N. E. F., French, J. M., Heaviside, D., Julian, D. G., and Shaw, D. A. (1986b). Long-term intellectual dysfunction following coronary bypass graft surgery: a six months follow-up study. *Quarterly Journal of Medicine (New Series)* **62**, 259–68.

Sheperd, J. P., Qureshi, R., Preston, M. S., and Levers, B. G. H. (1990). Psychological distress after assaults and accidents. *British Medical Journal* **301**, 849–50.

Simanowitz, A. (1985). Standards, attitudes and accountability in the medical profession. *The Lancet* (ii), 546–47.

Tennen, H. and Affleck, G. (1990). Blaming others for threatening events. *Psychological Bulletin* **108**(2), 209–32.

Thomsen, V. (1984). Late outcome of very severe blunt head trauma: a 10–15 year second follow-up. *Journal of Neurology, Neurosurgery and Psychiatry* **47**, 260–68.

Vachon, M. L. S., Sheldon, A. R., Lancee, W. J., Lyall, W. A. L., Rogers, J., and Freeman, S. J. J. (1982). Correlates of enduring distress patterns following bereavement: social network, life situation and personality. *Psychological Medicine* **12**, 783–88.

Vincent, C. A., Pincus, T., and Scurr, J. H. (1993). *Patient's experience of medical accidents. Quality in Health Care* (in press).

Weiss, R. S. and Parkes, C. M. (1983). *Recovery from bereavement*. Basic Books, New York.

11

The effect of accidents and litigation on doctors

MAEVE ENNIS and J. GREDIS GRUDZINSKAS

A survey carried out in the USA in 1984 revealed the surprising statistic that 54.5 per cent of the responding doctors saw litigation as part of the day-to-day practice of medicine. They reported strong feelings of anger about this and many saw themselves as 'scapegoats for the legal profession'. They resented time being taken away from their clinical practice by involvement in the legal process and many felt that litigation eroded the quality of the traditional doctor–patient relationship. However, while many seemed to accept litigation as an inevitable part of medical practice, few saw it as an affront to their clinical competence (Charles *et al.*, 1984). That study and others like it have demonstrated that medical accidents and litigation can have profound effects on doctors both personally and professionally and on relationships with patients and patient care.

It has been estimated that in the USA at least one in four doctors will be sued for malpractice each year and where a doctor practices in a high–risk speciality the probability that he or she will be sued increases (Charles *et al.*, 1984). In the UK the figures are lower. The Medical Protection Society has 11 000 members, of whom approximately 1 in 10 write to the society each year. Many of their letters will be no more than questions and queries, but 25 per cent of them will be because a complaint or claim for compensation has been made against the doctor. This implies that 2.5 per cent of their membership has a claim or complaint against them in any one year (Palmer, Personal Communication). The majority of claims arise from relatively minor causes, and all specialties attract claims with some, such as obstetrics and gynaecology, accident and emergency, anaesthetics, and orthopaedics, attracting more claims than others (Ham *et al.*, 1988). Very few claims actually reach court: 70–75 per cent are dropped and of those remaining, 90–95 per cent are settled before they reach court (Fenn, 1992).

The fact that such a small proportion of cases reach court does not minimize the effect that litigation has on doctors. Many state that cases settled out of court will sometimes have a greater effect on them: they consider that the accident was unavoidable, since risks are inherent in particular procedures, or that they were not to blame and settlements out of court leave them with no chance to vindicate themselves. They may feel that they are left with a cloud hanging over them.

The effect of litigation or perceived risk of litigation

In the USA, studies suggest that some doctors are refusing to accept patients they regard as 'high risk', are retiring earlier, are leaving certain fields (e.g. obstetrics), or are ordering more tests than are medically necessary so as to defend themselves in the event of a lawsuit. Some also report discouraging their children from entering medicine (Sloan *et al.*, 1989; Charles *et al.*, 1984).

The most often-cited effect of litigation on clinical practice is the adoption of 'defensive medicine'. This is said to occur either when specific procedures are employed explicitly for the purposes of averting any possible lawsuit or by providing appropriate documentation that a wide range of tests and treatments have been used. Tancredi and Barondess (1978) have suggested that defensive medicine can be characterized as either positive or negative. Positive defensive medicine is the use of diagnostic or therapeutic measures to protect the doctor from being found liable. Usually these are measures that are unnecessary for the proper care of the patient. Negative defensive medicine refers to the withholding of procedures that might be medically justified in view of the patient's physical condition, but have in them an inherent risk of an adverse outcome and could thus be the basis of litigation. Positive defensive medicine may expose patients to the risks of adverse outcomes from the procedures themselves. Negative defensive medicine may result in suboptimal care by denying patients potentially beneficial procedures (Tancredi and Barondess, 1978).

Studies in the USA in which doctors were asked to report on their own practice provide most of the evidence on the extent of defensive medicine (Wyckoff, 1961; Tancredi and Barondess, 1978; Zuckerman, 1984; American College of Obstetrics and Gynaecology, 1985; Charles *et al.*, 1985; Weisman *et al.*, 1989). The study by Charles *et al.* (1985) in which physicians self-reported their reactions to litigation found that as a result of being sued the majority (69 per cent) report keeping more meticulous records but many reported recording less pertinent information. Many (62 per cent) ordered diagnostic tests for 'protection', even when clinical judgement deemed them unnecessary. Forty-one per cent stopped seeing 'certain kinds of patients' although they did not specify what kind of patient these were, and 28 per cent reported that they had stopped carrying out certain procedures, the failure of which may lead to lawsuit. A further study carried out by the same team looked at the reaction to litigation of both sued and non-sued physicians. 48.9 per cent of sued physicians reported that they were likely to stop seeing patients with whom the risk of litigation seemed greater. They also reported thinking of retiring early and discouraging their children from pursuing a career in medicine. However, the same behaviours were reported by non-sued physicians who had been asked to report on the threat of being sued. Both groups, in almost equal numbers, kept more meticulous records (74.5 per cent and 78.6 per cent) ordered more diagnostic tests that their clinical judgement deemed unnecessary (67.6 per cent and 59.6 per cent), and stopped performing certain high-risk procedures (42.8 per cent and 36.2 per cent). Many in both groups reported including information of little pertinence on patients' records. This suggests that a perceived threat of litigation has almost as much impact on clinical practice as an actual lawsuit.

The Harvard Medical Practice Study also asked physicians to report on their views of the perceived risk of being sued and the effect of that risk on their practice. They found that overall the perceived risk of being sued in any one year is 19.5 per cent, approximately three times the actual risk of being sued. When negligent care had been delivered the perceived risk of being sued rose to 60 per cent, a figure much higher than the actual risk of litigation from injuries caused by negligence. Also they found that the perceived risk of litigation was significantly higher for high-risk specialties such as obstetrics or orthopaedics.

In the Harvard study, perceived risk appeared to have the effect of requesting more tests and procedures (which may or may not be appropriate). Respondents who reported higher perceived risk also reported that they reduced the scope of their practice more often than those with lower perceived risk of being sued (Harvard Medical Practice Study, 1990).

Among medical specialties, the most frequently sued doctors both in the UK and the USA are obstetricians. Seventy-three per cent of the 24 500 members of the American College of Obstetricians and Gynaecologists have been sued for malpractice and it has been claimed that 3000 have abandoned the speciality to escape soaring insurance costs (Schreiber, 1987). In the USA, obstetricians and family physicians increasingly report that they are eliminating the obstetric portion of their practice or reducing the provision of care to patients who are identifiably at high risk because they fear being sued and do not want to accept the high cost of liability insurance. Overall, the American College of Obstetricians and Gynaecologists found that in 1987, 27 per cent of its members had reduced or eliminated the provision of high-risk care. A study carried out in Washington State examined all physicians who purchased insurance from one particular company between 1982 and 1988. Of the 690 doctors studied, 171 (32 per cent of the family physicians and 10 per cent of the obstetricians) discontinued obstetrics but remained in practice. The obstetricians who dropped obstetric practice had over twice the number of claims against them than their colleagues who remained active in obstetrics. Of the family physicians who stopped practising obstetrics, age, not new claims, was the most significant factor, but the authors of the study feel that it was the threat of malpractice claims in these older doctors that led them to stop delivering babies. They suggest that economic and lifestyle considerations are probably as important as litigation issues in the decision to discontinue obstetrics. Also these doctors may be concerned that continuing obstetric practice may expose them to potential malpractice suits even after retirement (Rosenblat *et al.*, 1990). This conclusion was supported by the fact that the older doctors retiring from obstetric practice had fewer new claims than younger colleagues who remained obstetrically active. But this could be a factor in itself. Observing the significant emotional and practice disruption in a colleague who has been sued may influence the decision to discontinue obstetric practice.

However, a note of caution must be inserted here. There are well-recognized biases involved in self-reporting in general and particularly in studies such as the above-cited study carried out by the American College of Obstetricians and Gynaecologists in which the respondents were asked to comment on salary, on malpractice, and on the threat of malpractice. This may perhaps have prejudiced their response in that direction.

It has been suggested that in the UK obstetricians have 'joined the ranks of the big spenders in medical litigation and it is going to get worse' and that the impact of this on practice is a more defensive attitude on the part of doctors ('more investigations, more operative intervention, and less opportunity to test the efficacy of existing methods of management' (Symonds, 1985)). Other authors have suggested that the rise in Caesarean section rate and investigations may in part be attributed to fear of litigation (Halle, 1986), that litigation or fear of litigation hinders recruitment to the speciality and encourages the practice of defensive medicine (Simmons, 1990). A survey by the Maternity Alliance into reasons for increased incidence of Caesarean section found that 28 per cent (38 doctors) of their sample cited litigation or fear of litigation as a factor (Boyd and Francome, 1986).

More precise information on the subject of defensive medicine in the UK is, however, hard to obtain. Our own investigation in this area suggested that it is being practised by limited numbers of obstetricians in Great Britain and the Republic of Ireland. We surveyed all fellows and members of the Royal College of Obstetricians and Gynaecologists in the UK and Eire on the current state of practice. We asked respondents to indicate their attitudes to the accuracy of the most commonly used tests of fetal surveillance and if they continued to use the test despite finding it slightly, moderately, or very inaccurate or were unsure of this accuracy (Ennis *et al.*, 1991; see Table 11.1). We also asked for reasons for continuing the use of this test in these circumstances. We found that the most frequently given reason for continued use of tests perceived to be inaccurate was 'aid to clinical judgement' which was scored positively by 76 per cent of continuing users of all tests except

Table 11.1 Reasons for continued use of fetal surveillance tests deemed to be inaccurate (from Ennis et al., 1991)

	Ultrasound (%)	CTG			Fetal blood sampling (%)	FMC (%)	Biochemical tests (%)	Doppler blood flow (%)
		Ante-partum (%)	Intra-partum					
			Intermittent (%)	Continuous (%)				
Inaccurate but still used	90.2	81.6	70.1	100	60.6	86.9	23.2	84.3
Aid to clinical judgement	92‡	89.8‡	76.5‡	87.9†	21.6	82.5‡	14	68.9*
For the medical record	21.1	43.6	47.6	45.8	21.6	27.4	14	14.8
Medico-legal reasons	24.6	51.8	47.3	60.8	59.5‡	21	25.5*	8.1
Unit protocol	2.3	14.9	19.4	24.4	24.3	13.3	25.5	49.7

* $p < 0.05$; † $p < 0.005$; ‡ $p < 0.001$. Statistically significant differences are shown between the 'aid to clinical judgement' and 'medico-legal' groups

Doppler blood flow (68.9 per cent), fetal blood sampling (21.6 per cent), and biochemical tests (14 per cent). But many respondents also cited medico-legal reasons for retention of fetal blood sampling (59.9 per cent) and biochemical tests (25.5 per cent), despite the fact that biochemical tests were seen as least accurate by all respondents. Medico-legal reasons were also scored positively by 47.3–60.8 per cent for intra-partum tests and 21–24.6 per cent for ante-partum tests except for Doppler blood flow studies (8.1 per cent).

There were a further 14.8 per cent who used the tests 'for the purpose of medical record' and 49.7 per cent for 'adherence to unit protocol', which we feel may be a more subtle expression of concern about medical litigation (Ennis *et al.*, 1991). These findings suggest that defensive medicine is being practised to a limited extent by obstetricians in the British Isles. Some may argue that keeping better records and adhering to unit protocols can only improve standards of care. This may be true but when defensive medicine also includes administering tests that the practitioner see as inaccurate or of limited accuracy, then we are looking not at improved standards of care but anxiety on the part of doctors which may lead to intervention. These interventions, for example, induction of labour or Caesarean section, although well-intentioned may in themselves be the cause of risk to mother or fetus, further adding to the stresses of practice in what many obstetricians see as a highly litigatious field.

The effects of accidents

The effects of medical accidents as opposed to medical litigation has been little reported in the literature. One such study carried out by Wu *et al.* (1991) in the USA looked at whether or not junior doctors learned from their mistakes, that is, whether mistakes change clinical practice. In this study, a questionnaire was sent to 254 interns working in three different hospitals. The questionnaire asked the respondents to describe the most significant mistake in patient care they had made in the last year which had serious or potentially serious consequences for the patient and which would have been judged as such by knowledgeable peers. Their findings were that junior doctors had very mixed responses to medical accidents for which they felt responsible. The most frequently reported types of error were missed diagnosis (30 per cent), errors in prescribing and dosages of drugs (29 per cent), and errors in evaluation and treatment (21 per cent). In response to the question 'what was the outcome of the mistake?', 90 per cent reported that there were severe adverse outcomes for the patients following the mistake and in 31 per cent of cases the outcome was death. Most respondents reported constructive changes to their practice following the accident. These were seeking more advice (62 per cent), increased vigilance (82 per cent), personally confirming data (72 per cent), and trusting others' judgement less

(49 per cent). Much smaller numbers reported defensive changes. Keeping mistakes to oneself (13 per cent) and avoiding similar patients (6 per cent) occurred less frequently. Whether respondents reported ordering more tests or other defensive procedures was not stated.

We replicated the Wu study and found that the outcome for patients was not so severe but defensive response was greater with 18 per cent reporting 'keeping the mistake to oneself' and 18 per cent reporting 'avoiding similar patients'. However, we also had 34 per cent who reported they were more likely to order more diagnostic tests as a result of the mistake (Ennis, un-published data). One extreme example of this was a report from a respondent working in an Accident & Emergency Department in which he described a case of a very elderly patient who presented several days after sustaining a burn to his foot because the wound would not heal. The doctor prescribed topical antibiotics and dressings and discharged the patient with no follow-up appointment. Unfortunately the patient had an underlying diabetic condition which he failed to disclose to the doctor. He did not seek further medical help for some months and eventually microangiopathy led to gangrene and the loss of the foot. The doctor reported that in response to this mistake he is in general more likely to order more diagnostic tests and that in particular he orders a urine test for every patient he sees regardless of presenting symptoms.

A further study carried out by Christensen *et al.* (1992) also looked at the effect of mistakes on physicians. This study used an interview format to probe how physicians felt about their mistakes, how their manner of coping with the mistake influenced their emotional responses, and the changes in their practice as a result of the mistakes.

The numbers in this study were very small (only eleven doctors were interviewed), and one doctor felt that he had not made any mistake in many years of practice, although he reported there had definitely been some 'bad outcomes' in his patients. However, although the numbers are small, this study is interesting in that it looks at the responses of experienced doctors, who had been in practice for many years at the time of the study (range 4–18 years) and who had some responsibility for the training of junior doctors. The findings of this study were similar to the studies cited above, despite the greater experience of the participants. Again, the most frequently reported type of error was missed diagnosis and medication errors. The outcome for patients was similar, with four patient deaths reported. None of the mistakes had resulted in litigation or threat of litigation. Ten of the eleven doctors interviewed reported that they had learned from their mistakes and had improved their practice in some way. Changes included modifications in the way diagnostic and treatment procedures were carried out, changes in the teaching of junior staff, and changes in doctor/patient communication. These physicians also reported that they kept the mistake to themselves.

Although these studies are flawed in that they depend on hindsight report-ing, low response rates, and one had very small numbers, they do suggest that mistakes and errors, even when they are undetected by others, and do not result in litigation, do affect subsequent medical practice both positively and negatively.

The effect on doctors personally

According to Lavery (1988),

Most physician/patient relationships begin when an apprehensive patient approaches a physician for cure. The relationship evolves and becomes stronger as the physician shows concern and provides support. The physician may unwittingly foster the view that we live in a land of miracle cures and thus instill unreal expectations. A malpractice suit is a hostile act that severs this close bond and destroys all trust. The desertion of a previously loyal member of the 'professional family' profoundly affects the practitioner. The response is anger because the loss is an adversarial event. By bringing legal action, the patient also assaults the physicians credibility, insinuating faulty judgement or treatment. Self-esteem and status as a successful practitioner in the community or member of the academic environment are suddenly jeopardized. A malpractice suit challenges professional reliability and authority. The physicians whole professional future hangs in the balance.

The anger and apparent sense of betrayal in the above paragraph is not unusual in doctors reporting on the personal effects of litigation. Doctors will tell of feeling 'utterly alone', 'isolated from colleagues and patients' (Charles and Kennedy, 1988) and of 'experiencing deep and visceral emotion' (Lavery, 1988). It has been said by Charles that the threat of malpractice compels the doctor to view his or her patient as a 'future adversary in a courtroom proceeding' (Charles, 1984).

Much of what we know of the personal effects of litigation on doctors is anecdotal, gleaned from interviews with doctors who have been through the litigation process, such as the studies carried out by Charles *et al.* (1984, 1985), or from doctors writing about their own experience (Lavery, 1988; Jackson, personal communication). The fact that these are personal and, therefore, subjective accounts does not obviate from the psychological distress reported by these doctors. Charles herself reported suffering insomnia, appetite change, irritability, headache, and a host of other symptoms characteristic of stress-induced illness. Charles, reporting on other doctors in the same situation, found that few doctors talked to anyone about the experience, reporting feelings of shame and isolation. Many take it as a personal attack, feeling that their personal integrity is under assault. They talk about losing confidence in their ability and losing the feeling of pleasure they previously had in their practice of medicine. This they feel has come about because litigation and threat of litigation subtly change doctors' relationships with all patients and not just those who actually initiate claims against them. As Charles has pointed out, there are doctors who view all patients as potential litigants, and this must affect the way they interact with patients (Charles, 1984). This must in turn affect the manner in which patients relate to their doctors. It has been suggested that many doctors are locked into self-important authoritarian professional roles which lay people rarely let them drop (McCue, 1982). McCue feels that a doctor is always expected to place a patient's interests before his or her own, that society enjoys seeing a doctor fail to meet its high standards, and likes to see a doctor who reveals vulnerability. This, he thinks, leads to doctors never being able to relax or fail or to escape the public eye or censure. This results in isolation which few other professionals experience when dealing with the public. This isolation has also been reported by non-sued physicians and it has been suggested that it extends to social situations as well as professional ones (McCue, 1982). Few doctors look for or receive any psychological or psychiatric help during or after a court case. Indeed many authors report a resistance on the part of most doctors, sued or not, to seek help even when depression or other psychiatric symptoms has been identified by the doctors themselves despite the fact that the present–day practice of medicine is intrinsically stressful (Roeske *et al.*, 1986; Thomas, 1976; Valko and Clayton, 1975). In Valko and Clayton's study of depression in interns only six of sixteen depressed interns saw their depression as abnormal. The equating of psychiatric or psychological symptoms with a sign of weakness was a common finding of all these authors.

Lavery (1988) drew an analogy between the doctors' reactions to a malpractice suit and the stages of grief reaction reported by those who have lost a close family member described in the literature on bereavement. He describes his own emotional reaction to litigation as being equally intense and suggests that understanding the psychodynamics of this reaction can help doctors to cope with the problems inherent in a malpractice suit. However, while he quite powerfully describes this psychodynamic model and its parallels with a malpractice case, his advice for coping with it is not to seek professional help (other than from lawyers), but to accept what has happened with 'humility and understanding towards the plaintiff' and use the confidence and assurance provided by other close relationships as support. This way he feels that those involved in litigation will better understand themselves and the process they have been through. But he fails to recognize that this denial of the need for professional help may itself be symptomatic of a maladaptive psychological reaction to the litigation process.

Who is to blame for litigation?

Many different views as to the causes of litigation have been put forward by doctors. Some doctors blame the profession itself, feeling that patients have been falsely led into unreal expectations of what medicine can and cannot do. Many feel that this is especially true in obstetrics: that patients believe that anything less than a perfect baby must mean that someone somewhere must have done something wrong. Lawyers have also been blamed: the American system of contingency fees and the sheer density of lawyers are often cited as the reason for the high rates of litigation in USA. However, Danzon (1986) did not find this to be justified and, in the UK, the difficulty of finding lawyers experienced in medical negligence has been identified as a serious obstacle to patients pursuing claims against doctors or hospitals (Capstick, 1990).

The media is also seen as contributing to the problem. Some doctors feel that their public image has been distorted by the media, in which reporting of cases that come to court is one-sided and often antagonistic (Jackson, personal communication). But there are also many who blame the public. Charles, for example, writes that 'the public must reflect on their attitude about compensation and decide whether every misfortune of life is compensable. If so, then how can society support that attitude? Until that occurs there is not going to be any solution. Only when the public becomes aware that they are the real victims of the medical malpractice crisis will change occur. Legislature changes will not happen until the public demands such changes. When the patient — with the aid of the medical practitioner — comes to a better understanding of the nature of the malpractice problem, its impact on physicians, and ultimately, its effect on health care, will the current attitudes towards initiation of malpractice suits diminish' (Charles, 1984).

Weinstein (1988) believes that the 'current mind-set of the public is responsible for much of the increase in the number of legal actions' and that the public has unreal expectations of doctors and medicine and that they treat every adverse outcome as an act of negligence. 'Modern man seems to lack any responsibility for himself', according to Weinstein. He also believes that a gambling spirit is inherent in the American psyche, that many Americans have a 'lottery mentality', and that the courtroom represents a lottery which affords that small chance to become a millionaire, without hurting anyone, since it appears to the public that no one individual is required to pay these large sums. However, unlike other authors cited, Weinstein acknowledges that medical malpractice does exist and may be contributing to litigation but he appears to regard it as a small part of what is going on despite describing the findings of a closed claims study as 'professionally embarrassing and unquestionably true' (Weinstein, 1988).

Why are doctors looking outside of themselves for causes of litigation? Many cite the very small numbers who are actually found at fault when cases do come to court as vindication for this attitude. But what they appear to be missing is that insurance companies in the USA and the defence societies in the UK only take to court those cases they feel they have a more than reasonable chance of winning. Most cases are settled out of court and while some of these cases may be settled because they are more of nuisance value than indications of justifiable claims, many will be cases that are indefensible. Closed claim studies and the Harvard Medical Practice study support this. While the numbers may be small there are patients who bring claims against doctors because their injuries are iatrogenic. The failure of doctors to recognize this may be one more symptom of the psychological effect of litigation and accidents on doctors.

The effect of blaming others for one's misfortune is well documented in the psychological literature, although studies have not looked at the behaviour of doctors. However, work by Tennen and Affleck, which reviewed a large number of studies, suggests that blaming others for threatening events is maladaptive and associated with impairments in physical and psychological well-being. Some of the studies they looked at found that blaming others was associated with depression, unhappiness, mood disturbances, poorer physical health, and less responsiveness to others (Tennen and Affleck, 1990). The association between blaming others and impairment was demonstrated by some of the doctors in our study of learning from mistakes. In this study we asked respondents to give comments on further effects of the mistake not covered by the questionnaire. While many confined themselves to comments on institutional responses to the mistake, some detailed how the mistake had changed the way they interact with patients and some of these changes suggest that some of these doctors are reacting in an inappropriate manner. For example, one consultant in psychiatry detailed a case in which a patient he had been treating committed suicide. The patient had made several previous suicide attempts and was being treated on an inpatient basis with antidepressants. Prior to going on holiday the doctor discharged the patient and reduced the medication. While the doctor was on holiday the patient committed suicide. The doctor comments: 'because patients like this are not to be trusted I know that I can never again go on holiday with my family and since this incident I have always been on 24 hour call' (Ennis, unpublished). This reaction is consistent with cognitive theories of depression that include unwarranted generalization as one of the perceptual distortions influencing negative self-evaluation (Peterson and Seligman 1985).

Responses to incidents like the one cited suggest that there are high levels of distress among doctors who experience accidents and litigation, and it has been suggested that the current trend towards promoting consumer rights, including the increasing use of litigation, will increase the pressure on doctors (BMA, 1992).

What's to be done?

Although the literature on the effects of accidents and litigation on doctors is sparse, there is enough evidence to suggest that the effects both on clinical practice and on doctors, personal lives is significant. Working in a profession which is intrinsically stressful, the added stressors of litigation and accidents is, not surprisingly, leading to maladaptive and inappropriate behaviours on the part of some doctors. The failure of doctors to acknowledge their need for help and to seek help when this need is recognized is part of the problem. There would appear to be a 'macho' attitude within the profession that admitting to suffering from stress and a failure to cope with the adverse effects of stress is a sign of weakness. If there is to be help for doctors who have been involved in litigation and accidents then there must be a change in this attitude. Firth-Cozens has found that doctors are often ill-prepared for the roles which they take on in the health service and suggests that both undergraduate and postgraduate education should play a role in teaching doctors to cope with stress (Firth-Cozens, 1987). The medical profession is one where risks and failures are inevitable and doctors must learn to deal with blame when things go wrong. This should also be a part of medical education.

A welcome statutory newcomer to clinical practice in UK is Clinical Audit in all disciplines, an activity which is directed at in-service training of practitioners at all levels of training to assess and discuss the consequences of their practice. Such activity will, at the level of undergraduate medical education, enhance awareness of the need for accuracy and precision when addressing any clinical situation, be it diagnostic or therapeutic. Awareness of where the scientific basis of clinical practice stops and the 'art of medicine' begins will lead to a stronger foundation of medical knowledge to enable tomorrow's medical practitioners to deal with the expectations of patients. The BMA suggests that the teaching of basics of medical law would help doctors understand the legal process which may help alleviate the stress and fear associated with litigation (BMA, 1992) (though where this would be fitted in to an already overcrowded curriculum is unclear).

Hospital management also needs to be educated to understand how stressful the day-to-day practice of medicine can be and to be more open with both doctors and patients when things go wrong. It should be up to management to promote open dialogue in what is often a confrontational situation between doctors and patients, resulting in a great deal of stress to both and much unnecessary litigation. Better communication in general is needed. Many patients say they resort to litigation because it is the only way they can find out what went wrong (Vincent *et al.*, 1991). Until a more open attitude is taken, we cannot expect to reduce litigation and its often devastating effects on doctors, their clinical practice, and their patients.

References

American College of Obstetrics and Gynaecology (1985). *Professional liability insurance and its effects: report of a survey of ACOG's membership*. ACOG, Washington, DC.

BMA (1992). *Stress and the medical profession*. British Medical Association, London.

Boyd, C. and Francome, C. (1986). *One birth in nine*. Maternity Alliance, London.

Charles, S. C. (1984). A different view of malpractice. *Chicago Medicine* 87, 338–42.

Charles, S. C. and Kennedy, E. (1988). *Defendant: a psychiatrist on trial for medical malpractice*. Free Press, New York.

Charles, S. C., Wilbert, J. R., and Kennedy, E. (1984). Physicians self-report of reactions to medical malpractice litigation. *American Journal of Psychiatry* 141, 563–65.

Charles, S. C., Wilbert, J. R., and Franke, K. J. (1985). Sued and non-sued physicians' self-reported reactions to malpractice litigation. *American Journal of Psychiatry* 142, 437–40.

Christensen, J. F., Levinson, W., and Dunn, P. M. (1992). The impact of perceived mistakes on physicians. *Journal of General Internal Medicine* 7, 424–31.

Ennis, M., Clark, A., and Grudzinskas, J. G. (1991). Change in obstetric practice in response to fear of litigation in the British Isles. *Lancet* 338, 616–18.

Fenn, P. (1992). The cost of the current medical negligence system and proposed alternatives in the UK. Paper presented at Conference on *Alternative system for compensating the victims of medical accidents: international experience and policy implications*. Centre for Socio-Legal Studies, Oxford.

Firth-Cozens, J. (1992). Emotional distress in junior house officers. *British Medical Journal* 295, 533–36.

Halle, H. M. (1986). The cost of medical defence. *Journal of the MDU,*

Ham, C., Dingwall, R., Fenn, P., and Harris, D. (1988). *Medical negligence: compensation and accountability*, p. 12. King's Fund Institute/Centre for Socio-Legal Studies, London/Oxford.

HMPS (1990). *Patients, doctors and lawyers: medical injury, malpractice litigation and patient compensation in New York*. Report of the Harvard Medical Practice Study.

Lavery, P. J. (1988). The physicians reaction to a malpractice suit. *Obstetrics and Gynaecology* 70, 138–41.

McCue, J. D. (1982). The effects of stress on physicians and their medical practice. *New England Journal of Medicine* 306, 458–63.

Peterson, C. and Selingman, M. E. P. (1985). The learned helplessness model of depression: current status of theory and research. In *Handbook of depression: treatment, assessment and research* (eds. E. E. Beckham and W. R. Leber), pp. 914–39. Dorsey, Homewood, Illinois.

Roeske, N. C. A., Clair, D., and Bittain, H. (1986). *A study of anxiety/depression, and adaptive behaviour of medical students, house staff and spouse/partners*. Paper presented at the annual meeting of the American College of Psychiatrists.

Rosenblatt, R. A., Weitkamp, G., Lloyd, M., Schafer, B., Winterscheid, L. C., and Hart, L. G. (1990). Why do physicians stop practising obstetrics? The impact of malpractice claims. *Obstetrics & Gynaecology* 76, 245–50.

Schreiber, S. C. (1987). Stress in physicians. In *Stress in health professionals* (eds. Roy Payne and Jenny Firth-Cozens). Wiley, Chichester.

Simmons, S. C. (1990). Compensation for damage at birth (Letter). *The Times*, 13 December.

Sloan, F., Mergenhagen, P., Bradley Burfield, W., Randall, R., Bovbjerg, J. D., and Hassan, M. (1989). Medical malpractice experience of physicians: predictable on haphazard. *Journal of the American Medical Association* **262**, 3291–97.

Symonds, E. M. (1985). Litigation in obstetrics and gynaecology. *British Journal of Obstetrics and Gynaecology* **92**, 433–36.

Tancredi, L. and Barondess, J. (1978). The problem of defensive medicine. *Science* **200**, 879–93.

Tennen, H. and Affleck, G. (1990). Blaming others for threatening events. *Psychological Bulletin* **108**(2), 209–32.

Thomas, C. V. (1976). What becomes of medical students: the dark side. *Johns Hopkins Medical Journal* **138**, 185–95.

Valko, R. J. and Clayton, R. J. (1975). Depression in the internship. *Diseases of the Nervous System* **36**, 26–29.

Vincent, C. A., Martin, T., and Ennis, M. (1991). Obstetric accidents: the patient's perspective. *British Journal of Obstetrics and Gynaecology* **98**, 390–95.

Weinstein, L. (1988). Malpractice — the syndrome of the 80s. *Obstetrics and Gynaecology* **72**, 130–35.

Weisman, C. S., Morlock, L. L., Teitelbaum, M. A., Klassen, A. C., and Celentano, D. D. (1989). Practice changes in response to the malpractice litigation climate. Results of a Maryland physician survey. *Medical Care* **27**, 16–24.

Wu, A. W., Folkman, S., McPhee, J., and Lo, B. (1991). Do house officers learn from their mistakes? *Journal of the American Medical Association* **265**, 2089–94.

Wyckoff, R. L. (1961). The effects of a malpractice suit upon physicians in Connecticut. *Journal of the American Medical Association* **176**, 1096–1101.

Zuckerman, S. (1984). Medical malpractice: claims, legal costs and the practice of defensive medicine. *Health Affairs* **3**, 128–34.

12

Compensation and accountability: the present system

CHRISTOPHER ORR

Old East comes to me in the morning with letters and did give him a bottle of North Down ale which made the poor man almost drunk. This day the Duke of Gloucester dyed of the smallpox — by the great negligence of the doctors. (Samuel Pepys, 30 September 1660).

The above quotation — an early, but by no means the earliest recorded lay perception of shortcomings in medical treatment — sets the theme for this chapter. With the passage of time there has evolved a growing demand for perfection in an obviously imperfect world. When this demand remains unfulfilled in whole, or in part, patients express their dissatisfaction through the increasing number of avenues open to them. As will be seen, recourse to these avenues is increasing steadily and this chapter sets out to explore their adequacy and fairness to all parties and whether they meet the needs of the aggrieved patients whilst still permitting the vast majority of doctors to treat their patients not only in a manner acceptable to their peers but retaining their clinical freedom and, dare it be said, right to decide what is in the best interest of a given patient without fear of undue interference from third parties.

The general practice complaints procedures

A minority of the population of the UK will require treatment in the hospital setting, whereas most people will at some stage require the services of one of 40 000 or so general practitioners. Many such persons make use of GP services minimally and for genuine need; others, however, for clinical or other reasons, require such services in much larger measure, and experience shows that it is from this substantial cohort that most complaints are generated: some justified, the majority not. These complaints have been, or are, dealt with under the following regulations:

- NHS (Service Committees and Tribunals) regulations 1974;
- NHS (Service Committees and Tribunals) amendment regulations 1974;

- Further amended by statutory instruments 74/455, 74/907, 82/288, 85/39, and 87/455 and currently under the NHS (Service Committees and Tribunals) amendment regulations 1990;
- NHS Service and Tribunal Committee amended regulations 1 April 1992 (apply to complaints received on or after that date).

Family practitioners comprising GPs, general dental practitioners, opticians, and pharmacists provide services to the public within the NHS as independent contractors. These independent contracts are made with the Family Health Services Authority (FHSA), formerly known as the FPC (Family Practitioner Committee). The contracts spell out specifically the services the practitioners are required to provide, and which are collectively referred to as the terms of service. To make a complaint it is only possible for a patient to contend that the practitioner may have been in breach of these terms and it is only possible for the relevant service committee of the FHSA to investigate the complaint if it appears that a prima-facie case of breach exists. Readers should note that there is no power to formally investigate complaints citing, for instance, rudeness, poor attitude, or demeanour in a practitioner who has otherwise fulfilled his or her contractual obligations. Such complaints can be dealt with informally but even when upheld, the offending practitioner suffers no penalty and the FHSA can take no further steps. This informal mechanism can also be invoked, subject to the agreement of both parties, where a minor breach of the terms of service may have occurred. If the complainant cannot thereby be persuaded to accept that this is not the case or even if the respondent practitioner does accept the patient's grievance but the complainant wishes to pursue the matter further, the formal procedure which now follows can be set in train. It is pertinent now to set out the two paragraphs of the terms of service which are most frequently taken into account when patients complain. These are Paragraph 3, which states:

where a decision whether any, and if so what action is to be taken under these terms of service requires the exercise of professional judgement the doctor shall in reaching that decision not be expected to exercise a higher degree of skill, knowledge and care than general practitioners as a class may reasonably be expected to exercise.

and Paragraph 13:

subject to Paragraph 3 a doctor should render to his or her patients all necessary and appropriate medical services of the type usually provided by general medical practitioners. They shall do so at their premises, or if the condition of the patient so requires elsewhere in the practice area or at the place where the patient was residing when accepted by the doctor or, if the patient was on the list of a practice declared vacant when the doctor succeeded to the vacancy or at some other place where the doctor has agreed to visit and treat if the patient's condition so requires and has informed the patient and the committee (Authority) accordingly. The doctor shall not be required to visit and treat the patient at any other place. Such services include arrangements for referring patients as necessary to other services provided under the

Health Service Acts and giving advice to enable them to take advantage of local authority social services. Except in an emergency, this paragraph shall not impose an obligation on the doctor to provide maternity medical services unless he or she has undertaken to do so.

When a complaint is sent in writing to an FHSA the officer responsible for dealing with complaints (such officers have varying titles) will refer the complaint to the chairperson of the relevant service committee, who must decide whether the complaint appears to constitute a breach and whether or not it can be dealt with informally as previously described. If there is a perceived breach which requires formal investigation, the respondent doctor is invited to comment on the substance of the complaint which he or she normally does with the assistance of the medical defence organization (MDO) to which he or she subscribes. It is likely that the MDO will have been aware of many impending complaints before they are formalized as doctors are usually, though often needlessly, concerned about the effect upon their reputation and practice should a complaint be upheld and are not slow to seek advice and reassurance at the first whiff of patient dissatisfaction. The doctor's response is shown to the complainant and his or her comments upon that response are sought. When the documentation is complete, the chairperson reconsiders it and decides whether or not a hearing is required. If he or she declines to convene a hearing, the complainant may appeal this decision to the Secretary of State or those to whom that Minister has delegated this task (see below). The complainant also has the right to appeal a decision not to admit a complaint out of time (13 weeks following the incident complained of).

In practice, hearings are on the increase and refusals to admit complaints for any reasonable cause are becoming rare. This increase has recently been dramatically confirmed by commentators from the defence bodies. Readers will not be burdened by a description of the procedure adopted by the service committee but it is important to note that both parties may now be represented by a friend or advocate paid or not and who is not a practising lawyer, which will see an increasing role for the MDO's on the one hand and Community Health Councils (CHCs) on the other. The outcome of the hearing, which is an investigation and not a disciplinary proceeding, will be that the complaint is either upheld or rejected. If upheld, the practitioner may be cautioned to adhere more closely to his or her terms of service in the future or, much more commonly, the committee will recommend to the FHSA that a withholding be made from the practitioner's remuneration. Any withholding in excess of £500 must be ratified by the Secretary of State who not infrequently increases it! Both parties may appeal a decision adverse to them and consideration of these appeals at the time of writing (April 1992) has been delegated to the Family Health Services Appeals unit in North Yorkshire rather than being dealt with as formerly at the Department of Health in Central London. (NB: practitioners may appeal against either the level of withholding, the finding of a breach, or both.)

At an oral appeal which is essentially a rehearing, both parties may be represented by a barrister or solicitor and the Secretary of State has the power to order the FHSA to contribute to complainants' costs in mounting and attending the appeal, and as appropriate both parties may be required to contribute to legal costs incurred. Where a substantial withholding has been made and confirmed there is an increasing tendency for the General Medical Council (GMC) or other professional body to take note of such withholdings, the reason why they were made, and whether the conduct of the practitioner possibly amounted to serious professional misconduct subject to severe disciplinary sanctions if proven. This aspect will be considered further when the present and future role of the GMC is examined.

Are the procedures considered above fair? On balance it would seem so, although patients on occasion seem unable to understand why it is that a doctor is not automatically penalized for getting a diagnosis disastrously wrong despite taking steps to reach a diagnosis considered by their peers to be reasonable. Equally, doctors seem aggrieved if in their view they are unjustly found guilty of failing to refer timeously for a second opinion. They sometimes believe this has impugned their clinical acumen and encourages a misuse of the hospital services by over referrals. These worries are probably understandable and it is here that the encouragement and advice available from the MDOs has much to offer in repairing a shattered professional ego.

Both parties, it is believed, will now benefit from the improved level of representation available to them at the primary hearing. However, it should not be forgotten that in addition to the complaints procedure patients may elect to pursue a suit in negligence against their doctors, especially if the latter have been found in breach or are censured by the GMC. The recent publication of the Patient's Charter — the subject of sardonic comment from Harris Myles (1991) — should further enhance the ability of the patient to express and seek redress from their grievances. Recent reports of an expressed desire by a proportion of general medical practitioners to modify their contracts so as to obviate the necessity to provide 24-hour cover will surely, if granted, generate a plethora of complaints since in order to provide 24-hour patient coverage it will be necessary to rely increasingly upon locum and deputizing services who may or may not have access to patients' records, albeit that an argument has been put forward that any competent doctor should be able to recognize and treat a real emergency without being privy to a patient's previous history and treatment.

Community health councils

These councils were first set up in 1974 as a means whereby the interests of patients treated within the NHS could be taken into account. The statutory arrangements creating CHCs provided for their establishment in each District

Health Authority and hence there are approximately 200 CHCs now in existence. Councils are constituted as follows. One third of council membership (approximately 22) is nominated by voluntary organizations, one half by local authorities, and one sixth by the Regional Health Authority who pays and appoints the Committee secretary in consultation with the CHC itself. Recent amendments to the enabling legislation (1990) preclude directors of NHS Trusts from membership of CHCs.

CHCs enjoy considerable powers, including access to NHS premises and employed staff, but most importantly — from the standpoint of this chapter — CHCs, or more properly their secretaries, advise patients how to mount complaints against general practitioners and the hospital services. Despite efforts to simplify the various complaints procedures available, some patients still find these difficult to comprehend; here, the role of the CHC secretary in guiding patients through the regulations is invaluable and one which is likely to be increasingly played with the advent of the Patients' Charter. Ideally, CHC secretaries should be even-handed in their dealings and not encourage vexatious and sometimes frivolous complaints.

For the most part, this appears to be the case; though there are known to be pockets throughout the country where CHCs are seen to be very much 'anti doctor' and have taken upon themselves the roles of prosecutor, judge, and jury, which may not be in the best interests of those they are appointed to serve. The author has had dealings with CHCs at opposite ends of the spectrum. first, when employed by one of the Medical Defence bodies when he saw letters of complaint drafted by CHCs levelled against members of the defence body concerned. For the most part these complaints, justified or otherwise, were set out in a fair and articulate fashion avoiding exaggeration, though a small minority were not. The latter seemed to reflect the personality and politics of their authors rather than the bare facts of the instance giving rise to the complaint. Currently, the author is employed by a Regional Health Authority and has been charged *inter alia* with ensuring that the CHCs within the Region discharge their duties to patients in the complaints sphere equitably and ensuring that all information necessary to fully inform the CHC secretaries is made available to them so that they do not misconstrue the facts and circumstances surrounding a patient's grievance. It must be remembered that the more advanced stages of the hospital complaints procedures (see below) are a charge on public funds and should be brought into play only when the patient cannot be otherwise appeased. The same remarks apply to appeals by patients from service committee decisions, the success rate of which, as has been seen, is relatively low. There now follows summaries of the work of CHC nationally and locally so that readers can see in perspective the scope of work undertaken by the councils.

Tables 12.1 and 12.2 are derived from a recent briefing from the Association of Community Health Councils for England and Wales and the annual reports of local CHCs respectively.

Table 12.1   Written complaints for the financial year 1987/88 in England (by type of complaint and method of investigation)

	Hospital	Community	Identification and charging of overseas visitors
Type of complaint			
Wholly or partly clinical	12 520	894	0
Other	17 436	4136	0
Total number of written complaints	29 956	5030	4
Method of investigation			
By officers only	28 053	4711	3
Referred to RHA, DHA, or BG members			
Further investigation unnecessary*	940	206	1
Investigation carried out by RHA, DHA, or BG members	661	89	0
Investigation by formal independent Committee of Inquiry#	38	1	0
Investigation carried out by two independent consultants+	92	0	0
Method to be decided	172	23	0

* Further investigation unnecessary refers to complaints reported to the Authority (or an appropriate subcommittee) for decision as to further action, where the decision was that further investigation was unnecessary
# Independent Inquiry — the procedure in paragraph 7(111)(B) of HM(66)15
+ Investigation of complaints concerning clinical judgement (as set out in part iii of HC(81)5)

The hospital complaints procedure

The Hospital Complaints Procedure Act received Royal Assent in July 1985. The Act requires the Secretary of State to issue directions as to how the Act was to be operated by Health Authorities and how the workings of the Act and access to its provisions should be publicized. In 1986, Health Authorities were asked to comment on the proposals for handling of complaints and following receipt of these comments a circular was issued by the Department of Health in June 1988 (HC88-37) which set down the rules and practices to be followed, although many Health Authorities had reported that the procedures that they employed were very much in line with the provisions of the new Act and were derived from an earlier circular (HC815) since rendered obsolete by the introduction of HC88-37. It was further directed that the current procedures were to become operational by 29 July 1988 and were to remain in force until 31 May 1993.

Table 12.2   Written complaints for the financial year 1987/88 to the East Anglian RHA (by type of complaint and method of investigation)

	Hospital	Community	Identification and charging of overseas visitors
Type of complaint			
Wholly or partly clinical	329	13	0
Other	556	67	0
Total number of written complaints	885	80	0
Method of investigation			
By officers only	880	80	0
Referred to RHA, DHA, or BG members			
Further investigation unnecessary*	1	0	0
Investigation carried out by RHA, DHA, or BG members	0	0	0
Investigation by formal independent Committee of Inquiry#	0	0	0
Investigation carried out by two independent consultants+	0	0	0
Method to be decided	4	0	0

* Further investigation unnecessary refers to complaints reported to the Authority (or an appropriate subcommittee) for decision as to further action, where the decision was that further investigation was unnecessary
# Independent Inquiry — the procedure in paragraph 7(111)(B) of HM (66)15
+ Investigation of complaints concerning clinical judgement (as set out in part iii of HC(81)5)

The new Act clearly recognizes the need for a vehicle through which patients and indeed staff can channel their unease and is relatively 'user friendly'; it is regarded as being most important that no-one is seen to be deterred from making a legitimate complaint and complainants have the right to expect that their grievances will be dealt with expeditiously, sympathetically, and comprehensively.

If investigation of a given complaint reveals a remediable defect or defects in systems of patient care there is a valid expectation that these, where identified, will be rectified. Before embarking on a detailed examination of the procedures currently in use, the author wishes to emphasize that medical complaints can often be defused at an early stage by sympathetic communication with patients where things are thought 'to have gone wrong'. It is, moreover, manifestly unfair to patients and indeed to junior doctors to delegate this possibly distasteful exercise to registrars and house officers. Consultants must set aside ample time to allay anxieties and doubts early, with candour, and with the stamp of authority vested in their post. It is only when these efforts fail that the steps which follow will need to be taken. Authorities are required to designate an officer responsible for the receipt, investigation, and due processing of clinical complaints. In practice, this officer is often a unit general manager or his or her immediate deputy. When a complaint is made concerning treatment, full information regarding the incident complained of must be made available to the officer who in turn must themselves be readily available to complainants who may be patients, staff, or others acting on behalf of the complainants with, be it noted, the latter's knowledge and consent if he or she is deemed to be capable of giving it. Those acting for complainants will almost invariably be secretaries of Community Health Councils whose role has already been examined.

At an early stage, the designated officer should ascertain, if possible and with advice, if a given complaint is likely to result in legal action. If such a likelihood is perceived, the steps which follow are inappropriate, though some patients proceed through all the stages of the complaints procedure and still elect to sue, which must always remain their inalienable right. The officer must also confine the use of the procedures to be described to matters involved with clinical judgement and practice. Matters which give rise to criminal charges, disciplinary action, or allegations of physical abuse to patients are dealt with in other ways and are not appropriate for further consideration within the context of this chapter.

The stages of investigation of clinical complaints are as follows.

The first stage. The complaint is first brought to the attention of the consultant responsible for the care of the aggrieved patient, albeit in practice the consultant concerned will in all probability be aware of it, certainly if the approach earlier proposed has been followed. The consultant will discuss the substance of the complaint with other doctors who may have been involved and thereafter will offer to meet the patient, possibly with junior doctors present and sometimes, but not always, designated officers. Following this meeting the designated officer will write to the complainant summarizing the meeting and offer any necessary apologies where these may be appropriate. If the complainants remain dissatisfied, they are then asked to put their complaint into writing if it has not already been done and the matter then proceeds to the second stage.

The second stage. The District General Manager will inform the regional director of public health (formerly the RMO), who usually discusses the complaint with the consultant and certainly if the case record does not contain an adequate statement (which it should). The Director may then arrange to interview the complainant or arrange for this to be done by a member of his or her directorate, as is the case in the region in which the writer works and where he or she is charged with this responsibility. It is considered that this interview serves a vital purpose and, though optional, should not be omitted. Interviews are conducted at the Region or at a place of the complainant's choosing so as to emphasize that the Region is acting entirely independently and impartially in an effort to resolve the issues at stake.

In the author's experience about half the complaints investigated at the second stage are resolved and no further action is taken. However, if the above exercise is not successful the next stage may be offered and invoked where this offer is accepted. Before making the offer, assurances should be sought that legal action is not in contemplation though in fact there are many recorded instances in which, despite such assurances, litigation has followed, especially when the report of those who assessed the complaint at the third stage is an adverse one.

The third stage. The Director or deputy approaches the joint consultants committee at BMA House to nominate two independent assessors from outside the Region. At least one of these will be working in a hospital comparable to that in which the complaint arose. If more than one specialty is involved, one of the assessors should represent each specialty. The assessors are then fully briefed with the written complaint, complete copies of the complainant's medical records, and such other information as they may require. They will then arrange to meet the complainant either in their home or in premises provided by the District Health Authority. They will also meet the consultants and other doctors concerned though not together with the patient. The patient may be accompanied to their interview with the assessors by a 'friend' (CHC secretary) or their GP. When the interviews are concluded the assessors formulate an agreed report which is sent through the instructing Region to the District General Manager who in turn writes formally to the complainant setting out the findings of the assessors and drawing attention if appropriate to the implementation of any remedial action that the assessors might have recommended.

In writing to the complainant the DGM may seek advice from the Director as to how he or she might phrase the comments of the assessors to ensure they are understandable though undiluted. The final report of the assessors remains confidential though it is, on occasion, shown in its entirety to the consultant. At this point, the clinical complaints procedure is exhausted and no further appeal is possible except perhaps to the Health Service Commissioner or Ombudsman if there have been any administrative elements to the complaint or administrative delays in investigating it.

Is the procedure described above a fair one? As before, the answer must be in the affirmative. Provided that patients are convinced of the genuine desire to redress their grievances, to avoid repetition of errors, and to treat them with respect in an open, even-handed manner. The writer is unfortunately aware of instances where at least one Health Authority has attempted to use the third-stage procedure to pillory a consultant or his or her staff. This should be seen as unacceptable practice and one which has the potential to frustrate the very intentions of the 1985 Act. As hitherto mentioned it is likely that clinical complaints will continue to increase and it is possible that in 1993 the procedures now in use may be modified in the light of experience and to reflect the laudable and increasing desire of politicians to demystify medicine and increase accountability. Finally, a recent publication (Donaldson and Cavanagh, 1992) has seriously questioned the adequacy of the system described above, citing amongst their criticisms inordinate delays which are in any event inexcusable and supporting the introduction of lay persons into the complaints procedure which is likely to meet with stern resistance from the professional bodies, albeit that lay participation in FSHA service committee hearings is the accepted norm. This article is commended for further reading.

The health service commissioner

This post, more commonly referred to as The Ombudsman for England, Scotland, and Wales, was created by the NHS Reorganization Act of 1973. The holder of the post can only be removed from office by Parliament and enjoys very considerable powers to ensure that the administrative side of the NHS services is correctly and timeously managed. The Ombudsman is not, however, empowered to deal with patient complaints which he deems to be related to matters of clinical judgement, and neither can he undertake matters pertaining to perceived deficiencies in general practice or in the ophthalmic or pharmaceutical services for which separate mechanisms exist. When the Ombudsman undertakes a complaint he, or more usually his officers, interview all parties to the complaint and take statements from them. At times he will wish to see doctors who themselves may have fallen below the required administrative standards in matters such as efficient management of waiting lists, inadequate provision of information, and poor record-keeping by themselves or their juniors for whose actions they are responsible. Doctors who are to be interviewed by the Ombudsman's staff will, often as not, be advised by the defence bodies, representatives of which are usually allowed to attend interviews if the interviewee so desires. The purpose of this is to make certain that the interview does not stray beyond the remit considered above. Equally, there is an obvious duty to see that the complainant confines his grievances only to those which can be properly considered.

When the statements have been collected they are compiled into a report to which all parties have access for amending purposes before the document is finalized. On occasions the time given for amendment given is short — perhaps too short — but in practice this is an uncommon and not therefore a serious problem. The final report amended or otherwise is them submitted to the Secretary of State who will wish to be satisfied that any remediable deficiencies have been repaired and who may lay the Ombudsman's report before Parliament.

The Ombudsman appears to be particularly keen to expose avoidable administrative failures which have denied a patient his or her right to have a complaint properly investigated and if such administrative failures are identified, criticism is apt to be severe.

Earlier reference has been made to the obvious ability of the newly introduced Patient's Charter to open a floodgate of complaints which will have to be addressed as they occur, and it is certainly on the cards that doctors may be accused of failing to meet waiting list targets, albeit that the ability to do so may be to a greater or lesser extent resource-dictated. There is a thin red line in the waiting list management process between administrative lapses and clinical judgement, and doctors and their advisors may well find difficulty in keeping the Ombudsman on the right side of this line. It has been said that the Ombudsman may approach but not go through the door marked 'clinical judgement', though he may knock loudly upon it. The author speculates that this loud knocking will gain in volume and vehemence and that the door may well be breached in the near future. To finally drive home the growing role of the Ombudsman, there were 1176 patient complaints considered by him in 1991 compared to 990 in 1990.

The general medical council

The Medical Act of 1858 brought into being the General Council of Medical Education and Registration in the United Kingdom, commonly referred to as the GMC (Richardson, 1986). The prime purpose of the GMC was initially the protection of the public from the ministrations of unqualified practitioners; indeed, the census returns of 1841 suggested that one third of the 15 000 medical practitioners in England were unqualified! (Richardson, 1986). The GMC in its earlier years confined its activities largely to the establishment of the medical register and the upgrading and standardization of medical education which it continues to do, and only gradually did it turn its attention to examination of the conduct of practitioners which may or may not have brought about harm to their patients. It is with these latter activities that this chapter concerns itself: specifically, how the council currently approaches complaints of clinical incompetence rather than those associated with moral turpitude, important and newsworthy though these may be. The council receives well over a thousand complaints and enquiries per annum, each of which is carefully examined and the majority rejected or deemed to be without the jurisdiction of the council. After further screening, a small minority of the original approaches survive, some of which are relatively trivial and are dealt with by correspondence. The cases that remain must now be divided into those where a doctor's fitness to practice by reason of ill health or misuse of alcohol or drugs is the cause for complaint and those where a doctor may have been guilty of serious professional misconduct. Those in the former group are dealt with under procedures which are largely outside the scope of this text, though not infrequently a doctor's misuse of alcohol may instigate a complaint by a patient who he or she has treated.

The mechanisms for dealing with such problems appear on the whole to be satisfactory, and patients should be reassured and fortified by the knowledge that doctors whose addiction or illness may have harmed them or, more importantly, may harm others is dealt with in such a way as to rehabilitate and if possible cure them under supervision, and that conditions can be imposed on their practice until these aims are achieved. The Health Committee of the council also has the power in extreme cases to suspend a practitioner for up to one year, which totally removes, for that period, their ability to engage in any form of medical practice. There thus remains a relatively small number of cases which it is deemed proper that the Professional Conduct Committee of the council should consider whether the respondent doctor is guilty of serious professional misconduct. It is into this small subcategory that a similarly small number of cases referred by FHSAs or Health Authorities falls. In the author's experience only cases of extreme or repetitive neglect of clinical responsibility reach the Committee, and the outcome has by no means always been as drastic for the doctor as may have been forecast. It is likely that the number of such cases will increase for reasons touched upon previously and the adequacy and limitations of the present procedures may justifiably be called into question. The GMC, being a statutory body, cannot internally change its rules or conduct but it is known that there is a desire to introduce a lesser offence than that of serious professional misconduct coupled with an ability to order retraining under conditional registration and an ability to demand that the retraining has been satisfactorily completed. It now appears possible that these changes may well be commended at least in the foreseeable future. Changes such as these will surely provide patients with the assurance that the medical profession does police itself and that the interests of patients themselves rather than the sanction of doctors once again assumes primacy in the *raison d'être* of the GMC.

As this book goes to press it has been reported that the GMC, at its November meeting (1992), decided to endorse the principles of performance review embodied in a May consultation paper. Comments received on the content of the paper will be taken into account when drafting legislation to be submitted to the Privy Council in order to permit amendment of the 1992 Medical Act (Anon., 1992).

Medical negligence — the current system in the UK

At the time of writing a patient, or in the case of an infant his or her parents or guardians, may seek compensation for injury and its outcome brought about by and causally linked to negligence. Negligence in a medical sense has three elements, all or which must be fulfilled in order to bring a successful action. These are:

(1)   a duty of care which is established automatically at the first doctor–patient contact in a professional sense;

(2)   a failure to exercise that duty of care in a manner acceptable to a responsible body of professional opinion;

(3)   quantifiable damage must flow directly from the negligent act.

At present, the burden of proof rests with the plaintiff, who must prove to the court on the balance of probability that they came about their present condition as a result of acts or omissions unacceptable to a responsible body of professional opinion. Conversely, if the defendant can show that the mode of practice challenged was one which met with the approval of a responsible body of opinion, it will be open to the court to prefer that opinion. It will thus be apparent that success or failure for either party may depend crucially on the quality of expertise available to them: the author has little doubt that until very recently, plaintiffs have been disadvantaged in this regard, a point which will be expanded later. The possibility that the burden of proof may shift from the plaintiff to the defendant under proposed EEC legislation is considered in Chapter 14.

It is not proposed to go into great detail regarding the milestones leading from the bedside to the courtroom; however, a few generalizations may be appropriate. A would-be plaintiff believing that a cause of action exists consults a solicitor to act for him or her — it is essential that the chosen solicitor has an established track record in personal injury litigation, since expertise in other fields such as conveyancing does not necessarily imply equal excellence in other areas of legal practice! Over the last few years the author has seen the emergence of specialist firms who, supported by bodies such as AVMA (Association for the Victims of Medical Accidents), are able to provide plaintiffs with representation comparable to that available to defendants through defence bodies or more recently through Health Authorities or Trusts legal advisors in respect of claims arising in the hospital and community services sector which are the sources of the vast majority of UK claims and which fall to be dealt with under terms of so-called Crown indemnity introduced by HM89(34). Those acting for the plaintiff will, with their consent, seek voluntary pre-action disclosure of their client's case records under the provisions of Section 33 of the Supreme Court Act of 1981. Provided that the plaintiff's solicitors make out a cogent case for their request this is rarely refused, although there is often inevitable delay in the process pending receipt of clinicians' comments on the one hand and agreement in respect of administration fees by the plaintiff's solicitors and/or the legal aid board on the other. On the very rare occasion when disclosure is refused or unreasonably delayed, a court order compelling disclosure can be sought: the necessity for this step should be rare. The notes are then inspected by the solicitors, their client, and an expert in active clinical practice in the specialty concerning which the allegations of negligence may be made.

It is of paramount importance to select carefully experts for both sides who are surely entitled to equal levels of honest expertise. Lists of experts willing to act impartially are available from professional organizations such as the RCOG and of course other sources. The plaintiff's chosen expert will in due course advise if theirs is a case to mount, which probably only occurs in at most a quarter of cases. These latter cases are then referred to counsel in order to settle proceedings. In excess of 90 per cent of formalized claims are ultimately settled and of the remainder, only half are fought on liability. It is considered that the aim should be to recognize early those cases which are patently indefensible and that these should be settled with the minimum of ado and unnecessary expenditure. Doctors have expressed, and continue to express, the belief that settlement may be made on the grounds of financial expediency and possible detriment to their professional reputation. To date, this does not appear to be the case. Claims are settled only when significant vulnerability has been exposed or occasionally when a successful defence might be compromised by the unavailability of a vital witness. It can be said with confidence that no doctor's reputation will be best served during the course of a prolonged, yet unsuccessful and much reported, trial. Neither should it be forgotten that the trauma to the plaintiff of giving evidence in court can be considerable, notwithstanding the relatively relaxed manner in which civil trials are currently conducted. From the foregoing, it will have become apparent that the 'due process of law' may be prolonged for all concerned, especially when the outcome remains uncertain. Efforts in recent years to accelerate the process include the provision for simultaneous exchange of expert evidence between parties (Donaldson, 1987) and the provision for exchange of factual proofs of evidence provided by witnesses: also the extension of the financial jurisdiction of county courts so as to enable a substantial number of claims to be transferred from the High Court. This latter change may, however, have been based upon a false premise since the transfer might overload the county court lists and thus prolong delays more than the important and therefore more costly cases that come to trial in the High Court. A further possibility for speeding up the process now exists under the provisions of the Access to Medical Records Act (1990) which will, in due course, allow rapid access to medical records generated after 1 November 1991, though clearly it will take time for the advantages of this new access to filter down.

Is the present adversarial system satisfactory to all parties? In the author's opinion it is not and it must be but a matter of time before it is changed probably by public demand. The current delays in settling or bringing to trial medical negligence claims and the element of 'lottery' that currently exists cannot for much longer be condoned.

The possibility of introducing some form of no-fault compensation (no-cause compensation) with its advantages and disadvantages will be considered in Chapter 14. However, certain other initiatives have been raised and indeed the uncertainties for litigants and the length of time to taken to achieve results may conceivably be replaced by other systems as the anticipated harmonization within the EEC gains impetus. At the time of writing (April 1992) the incumbent Secretary of State for Health has introduced a discussion paper on the possibility that a form of arbitration between opposing though consenting parties might be offered as an alternative to tort litigation. The proposed procedure, though theoretically swift, suffers from a number of disadvantages which include no right of appeal and no obvious provision for the process to be applied to infant plaintiffs. The Secretary of State's paper has been circulated to Regional Health Authorities and other interested bodies whose comments were invited before the end of January 1992. The results of this circulation are awaited with interest though at present views personally canvassed have been on the whole pessimistic.

Finally, how should settlements or awards, granted by the courts or by way or arbitration, be paid? Until recently, the successful plaintiff received a lump sum to cover pain, suffering, and loss of amenities, together with provision for future needs, loss of earnings, and the like. The computation of these awards, particularly in respect of future needs, is, despite 'best efforts', fraught with difficulty, especially when life expectancy is taken into account — if the latter is over-calculated the award will be excessive and if under-calculated, insufficient. Recently, there has been a move towards structured settlements which involve an element of the lump sum by agreement between the parties being set aside to purchase annuities to cover future needs during the actual lifetime of the recipient, income from which has considerable tax advantages as opposed to the case where a portion of the lump sum were to be invested in, say, a building society where the income from which would attract income tax in the normal way. The Department of Health has not as yet expressed a formal view on this recent trend but logic dictates that such a view ought to be a favourable one. At least one specialist firm, in anticipation of an expansion of structures, is emerging as a leader in the field.

Readers may wish to ask what, if any, is the continuing role of the medical defence bodies in the present and future climate. Their role in claims management since the introduction of crown indemnity has sharply diminished though there is at least anecdotal evidence that departmental advice to draw upon over two centuries of accumulated experience is to some extent being followed. Although the defence bodies exist primarily for the protection of their members' interests, it is postulated that in the UK at least these interests coincide closely with those of their employers. It would be a matter for serious regret if Health Service managers and their advisors were to ignore the MDOs who can provide swift and sound advice which will enable claims to be concluded as equitably and economically as the present system permits. There exists a misconception that MDOs have traditionally ignored the interests of the injured patient in preference to those of their members. Several years of experience shows that this has not been the case, and patients and others should be discouraged from regarding the MDOs as being devotees of the hackneyed phrase 'the cover-up'. In short, no reasonable doctor would wish to see a patient genuinely injured as a result of a negligent medical mishap denied the timeous receipt of compensation where this is patently due to them.

Conclusion

This chapter has endeavoured to explore within limits the present systems whereby medical treatment may be called into question. Some of the perceived defects of the status quo have been identified and it is to be hoped that remedies will be found compatible with the maintenance and furtherance of the doctor–patient relationship which must surely be a *sine qua non* of modern and future medical practice.

References

Anon. (1992). *British Medical Journal* 305, 1230.

Brook Barnet, J. W. (1986). In *Oxford companion to medicine* (ed. Walton *et al.*), pp. 651–62. Oxford University Press.

Department of Health (1991). *The patient's charter*. HMSO, London.

DHSS (1981). *Circular HM 81.5*. DHSS, London.

DHSS (1988). *Circular HM 88.37*. DHSS, London.

Donaldson, L. J. and Cavanagh, J. (1992). Clinical complaints and their handling: a time for change. *Quality in Health Care* 1, 21–25.

Donaldson, M. R., Glidewell, L. J., Lawton, Sir F., Naylor, V., and Preston, H. A. (1987). *All England law reports*, pp. 353–367. Butterworth.

General Medical Council (1992). *Proposals for new performance procedures*, Consultation paper. GMC, London.

Ham, C. (1991). *The new National Health Service — organisation and management*. Radcliffe Medical Press, Oxford.

Myles, H. (1991, 26 November). What we need is a doctor's charter. *Daily Telegraph*, p. 13.

Richardson, The Lord (1986). In *Oxford companion to medicine* (ed. Walton *et al.*), pp. 435–43. Oxford University Press.

Waldegrave, Rt. Hon. W. (1991). *Arbitration for medical negligence in the NHS*. Consultation paper, Department of Health, 29 October 1991.

13

Compensation for medical injury: a review of policy options

PAUL FENN

Introduction

The preceding chapter in this volume has described fully the nature of the current system for compensating the victims of medical injury through the courts. In fact, it is often overlooked that victims are far more likely to be compensated by other means, and that total expenditure on compensating the medically injured through social security in any one year dwarfs the amount spent on tort damages.[1] This is clearly not a reflection of the relative generosity of the alternatives, but rather a reflection of the very small proportion of the medically injured who succeed in obtaining settlements under the tort system. The Harvard Medical Practice Study estimated a ratio of claims to negligent adverse events of 1:9 (Brennan, 1992), with less than half of these claims receiving payment.

Of course, it may not be appropriate to compare the tort system in this way with alternative compensatory mechanisms: the fault system may have other objectives besides compensation, including the preservation of medical accountability (Ham *et al.*, 1988). The latter objective is to be the concern of another chapter in this volume (see Chapter 14); this chapter will review the case for and against tort principally on compensatory grounds, and will similarly evaluate the proposed alternatives from this perspective. Nevertheless, the interdependence of compensation, accountability, and rehabilitation means that redistributive policy in this area can rarely be examined in isolation from its efficiency consequences.

The following section will evaluate the tort system as it currently stands; the next section will then suggest a 'blank sheet' framework for discussing the principles involved in compensation policy; a further section will discuss the criteria against which policy should be evaluated, and these will then be applied to particular options for reform.

---

[1] I have estimated the total cost of medical negligence claims to the NHS to be in the region of £60 million in 1990/91. This compares with over £5 billion for combined invalidity benefit/disablement benefit in the same year. It is impossible to estimate the proportion of total beneficiaries who were medically injured.

Tort as an imperfect compensation system

There are two viewpoints from which the tort system can be evaluated as a compensatory mechanism: looking backwards from the perspective of the successful claimant, or looking forwards from the perspective of the injured victim. Clearly, for those fortunate enough to succeed in securing damages through the tort system, it represents very generous compensation. The principle of restitution requires the courts to place the successful plaintiff in the position he or she would have been in, had the defendant not injured him or her, so far as this is possible through an award of monetary damages. This means that all economic loss resulting from past, present, and future interruptions to work as a result of the injury are incorporated into the award as well as any expenses involved in the care of the plaintiff. Also, some attempt is made to take into account the 'non-economic' losses associated with pain and suffering, loss of amenity, and so on. The total award is calculated as a lump sum, the returns from which investment would be sufficient to meet the objective of full restitution. In comparison with other compensatory mechanisms, this is generous to the successful claimant, although it is not beyond criticism. It has been argued that the lump-sum nature of the award makes an unwarranted assumption of financial prudence (and in fact structured settlements, by which payments are spread over the claimant's lifetime, are becoming increasingly common). It could also be said that the awards made for non-economic loss in English courts are in many ways symbolic only, and do not bear much relationship to the amount of compensation that the victim would see as equivalent to his or her loss. Moreover, the treatment of 'collateral benefits' — the provision of care and compensation from other sources — is said to be complex and inconsistent in many tort cases.

Notwithstanding these points, successful litigants will generally be satisfied with the generosity of the compensation they receive, if not with the way in which they had to obtain it. By contrast, potential litigants may not be so happy with the system as a means of securing compensation. It is from this perspective that tort has attracted its most severe critics. Essentially, it is argued that tort is flawed because it is *costly*, because it is subject to considerable *delay* between the accident and its compensation, and because it is *unfair*.

1.   *Cost*. The limited and declining availability of Legal Aid in civil cases is a serious barrier to the ability of many injured patients to pursue claims. Only the affluent and the indigent have open access to justice, while many people of modest means can only go to law at the risk of their entire personal capital, including their home.

2. *Delay.* At the extreme, a case might not come forward until 21 years after the alleged damage occurred and then take another 4 or 5 years to reach a conclusion. In the meantime the claimant is obliged to survive on alternative benefits to which he or she may be entitled.

3. *Fairness.* The tort system creates a fundamental inequity between injured patients, depending upon whether or not they can prove the causal connection between event and outcome, and the fault of a particular, identified, individual. Moreover, their success in doing so may depend upon the experience and training of the solicitor they choose. For each of these reasons like needs may receive unlike compensation.

Compensation policy reform in principle

A discussion of options for compensating medical injury essentially involves consideration of two questions: who should pay, and how much should be paid? In answering the first question, there are ultimately only three candidates: the doctor, the patient, or the state. The costs resulting from the injury, once it has occurred, must fall somewhere. If no one is held legally liable for these losses, the patient must bear them all. If doctors are held strictly liable for these losses, they bear all of the costs. If doctors are held liable only for negligently caused injuries, then some losses are borne by patients, and some by doctors, depending on the circumstances of the accident. Finally, the state may choose to bear some or all of the victim's losses through social security financed through general taxation or a levy on goods and services.

Once these institutional decisions about where losses should fall have been determined, it is then up to the parties concerned whether or not to make arrangements to spread, or pool, the losses through insurance. Obviously the state is large enough to self insure. The other parties, however, if risk-averse, will wish to buy insurance. Doctors may purchase third-party liability insurance against the cost of meeting patient claims. Alternatively, they could form mutual societies for the pooling of such losses, or negotiate a risk-spreading arrangement with their employers. Patients may purchase first-party insurance against the losses they have to bear themselves, either directly, by means of an income replacement or medical expenses policy with an insurance company, or indirectly, by means of a negotiated sick pay scheme through which employers meet such losses up to a maximum as part of a wages and conditions package.

Given these arrangements on liability and insurance, the determination of the amount of compensation depends on the way in which the victim's losses are calculated. To what extent should non-economic losses such as pain and suffering be included, and how should they be calculated? How should the future costs of medical care be estimated? What are the additional nonmedical costs associated with disability? These questions need to be addressed by the courts, insurers, or government agencies concerned with compensation.

The benefits and costs of policy reform

Given the array of options outlined above, we need a framework within which we can appraise any particular avenue of reform. What are the benefits, and what are the costs, associated with these arrangements?

Clearly, the benefits of shifting the incidence of the costs of medical accidents from one group to another will depend on considerations of efficiency and distributional equity: which group is it most socially beneficial to bear the loss, either because we wish that group to take the costs into account when deciding how to behave, or because we consider a particular group to be more 'deserving' in some way? The benefits of risk-sharing arrangements for whichever party bears the loss are evident: if that party is 'risk-averse', he or she will gain some satisfaction from an arrangement which permits the transfer of some of that risk to others who are more willing to accept it. Arrangements which permit risk-pooling where none was previously possible yield social benefits under this criterion, as would arrangements which transfer risk from relatively risk-averse people to relatively risk-neutral people.

As far as costs are concerned, there are essentially two categories: direct and indirect. The direct costs of transferring payments and pooling risks are those which result from the administration of these arrangements. Transferring wealth or risk from one person to another by itself does not involve a cost to society; it is the cost associated with implementing these transfers, determining eligibility, writing contracts, and so on, which is the real measure of the social burden of any particular set of compensation arrangements. Added to these, of course, are the costs borne by claimants as a result of delay in payment, and the associated anxiety, missed opportunities, and so on.

The indirect costs of compensation arrangements are those which result from the change in claimants' behaviour as a consequence of the receipt of benefits. For example, depending on the eligibility criteria adopted, and the rules for determining the level of benefits, some arrangements may affect the rehabilitation of injured beneficiaries more than others.

Appraising options

TORT REFORM

The first avenue of reform is to improve the efficiency of current liability arrangements; that is, to consider the extent to which the existing provisions for the compensation of the medically injured can be improved upon, without abandoning the principle of fault as the basis for eligibility. Movements in this direction have already taken place: the introduction of NHS indemnity in 1990 could be seen as an attempt to simplify the process of litigation by ensuring that there would normally be only one defendant — the relevant District Health Authority (Fenn and Dingwall, 1990). At the same time, following the recommendations of the Lord Chancellor's Review Body on Civil Justice, improvements have been introduced with a view to reducing court delays. Personal injury cases below £50 000 (which include the vast majority of medical negligence cases) now begin in the County Court; new rules have been introduced requiring earlier disclosure of medical reports; and improvements have been made to the arrangements for pre-trial exchange of witness statements (Civil Justice Review, 1988). Also, the Department of Health (1991) have recently proposed that an Arbitration scheme be set up in respect of claims for medical negligence against the NHS, with a view to reducing the costs and delays associated with litigation. Finally, the Green Paper on Reform of the English Legal System (Lord Chancellor's Department, 1989) suggested that some movement towards a contingent fee system may be appropriate, such that solicitors would agree to represent clients in exchange for a premium fee if successful (but unrelated to the actual amount of damages awarded). It could be argued that this would increase the chance of a non legally-aided plaintiff gaining access to legal services.

To the extent that the problems with the current system are the result of its direct costs in terms of legal and judicial resources, which impact upon access to compensation and administrative delay, then the range of improvements outlined above have some merit. However, because the fundamental allocation of costs would not change — that is, the burden of non-negligently caused accidents will remain with the patient — the problem of fairness will remain.

STRICT LIABILITY

One possibility for shifting some of the burden of injury away from the injured patient would be to change fundamentally the basis of liability, and remove the requirement that the fault of an individual be demonstrated before compensation is granted. It would then be possible to transfer the burden to those who have caused it, whether or not they were negligent in doing so. As explained above, this requires a shift from negligence liability to strict liability, with the medical profession made responsible for compensating losses resulting from their actions. However, because of the problems which would arise in the pooling of the increased risk amongst individual doctors, it might be preferable to restrict such liability to the organizations within which doctors are employed. This is essentially the proposal which Weiler and others have put forward in the USA (Weiler, 1991).

As a compensation scheme, this proposal ensures that a wider range of accident victims — that is, those who can prove causation but not fault — would be eligible for benefits. It also has the advantage that it can be financed by means of implicit insurance contracts between doctors and hospitals, with the latter passing on some of the cost in turn to the insured public (or taxpayers). Moreover, it has been argued that such a scheme carries with it an additional benefit through imposing the cost of accidents on organizational decision makers who are in a position to influence the activities of doctors within their institution, thereby preventing further accidents. The plausibility of this argument depends on the extent to which medical accidents are primarily a consequence of the range and type of new and complex medical interventions, and the management of associated resources, as opposed to the care undertaken by the individual physician.

In the UK context, the Weiler proposal is of particular interest, in that the introduction of NHS indemnity has put in place an administrative mechanism for dealing with negligence within the organization which could in principle be extended to a regime of strict organizational liability. The replacement of a negligence test with one of causation would change the emphasis of current procedures away from the identification of individual clinicians alleged to be at fault, towards an assessment of procedures which are associated with a greater risk of unanticipated injury to patients. Clearly, these organizational incentives will depend on the fact that the health authority or hospital itself bears at least some of the cost of compensation. If the organization is able to transfer all of the risk of claims to a higher tier of the health service through pooling, or through retrospective reimbursement, this would remove the need for management to control the risk. Consequently, it would be necessary to ensure that DHA or hospital budgets are set prospectively if the risk management benefits of strict organizational liability are to materialize. Similarly, if providers attempt to pass on the cost of claims to purchasers, the onus would be on the latter to insist on adequate risk management procedures as a contractual provision.

NO-FAULT

If the incentive benefits of strict liability are not deemed to be important, either because management will not respond or because other means of holding them accountable are believed to be more appropriate, then compensation based on cause may not be coupled with organizational liability. That is, government may accept responsibility for funding a cause-based scheme. This is what has come to be known as 'no-fault' compensation.

These may be divided into schemes limited to the adverse consequences of medical interventions only, of the type operating in Sweden and Finland and advocated by the British Medical Association; and general schemes, of the type operated in New Zealand, which happen also to include compensation for the victims of medical events.

Under the Swedish Patient Insurance Scheme, established in 1975, the county councils (which provide health services in Sweden) pay an annual premium per inhabitant to a consortium of private insurers who then manage claims and payments. The scheme covers treatment injuries, diagnostic injuries, incorrect diagnoses, accidental injuries, and iatrogenic infections or omission to treat. Victims retain the right to bring a tort action if they choose not to apply to the scheme for compensation.

Eligibility is restricted to injuries which cause more than 30 days' illness, 10 days' hospitalization, permanent disability, or death. The scheme costs about 70p per head of population per year, with about 16 per cent being paid out on administration and an average payment of £3200 to about 2500 people every year. However, it is important to recognize that this payment is essentially only a means of supplementing the relatively high levels of social insurance and health care provision in Sweden. In effect, it corresponds only to the 'pain and suffering' element in UK tort awards. Moreover, some 40 per cent of claimants receive no payment under the scheme. By retaining the test of causation, no-fault schemes will always raise equity problems, whether they are limited to particular categories of case, as in Sweden, or cover the whole population, as in New Zealand.

The New Zealand Accident Compensation Corporation (ACC) was set up in 1975, financed by special levies on employers and motorists and by general taxation. In the event of an accident, the victim can claim periodic payments of compensation for loss of earnings up to 80 per cent of previous income; lump-sum payments for permanent loss or impairment of a bodily function and for loss of enjoyment of life, pain, and suffering; and reimbursement of medical costs. It does not cover all injuries which result from omissions or failures of treatment or diagnosis or those which can be defined as routine complications. Moreover, inequities have emerged between those disabled by accidents and those whose disability results from a congenital cause, from sickness, or from ageing.

The British Medical Association has long been an advocate of a no-fault scheme for the UK (BMA, 1983). More recently, a Working Party set up by the Royal College of Physicians recommended a variant of no-fault for adverse medical events loosely based on the Swedish Patient Insurance scheme (Royal College of Physicians, 1990). In 1991, a similar proposal was presented to Parliament in the form of the NHS (Compensation) Bill, a private member's Bill introduced by Rosie Barnes MP. Despite a vigorous debate, the government remained unmoved, and the prospect of such a scheme being legislated in the UK seems remote.

What then are the basic features of the three no-fault schemes to have been proposed for the UK?

The BMA's scheme. Eligibility would be based on proof that the injury was related to the medical treatment received, and not merely the consequence of the progress of the disease under treatment. Excluded would be unforeseeable diagnostic error, unavoidable complications and infections, and complications of drug therapy carried out in accordance with the manufacturer's instructions. *Benefits* would comprise a capital sum reflecting conventional levels of compensation for pain and suffering, together with the periodical payment of compensation for economic loss, with income replacement limited to twice the national average wage. *Administration* would be through a 'No-Fault Compensation Board', and the scheme would be financed out of general taxation.

The Royal College of Physicians' scheme. Eligibility would be based on proof of 'adverse consequences of medical intervention'. *Benefits* would cover pain and suffering as well as economic loss, with all elements subject to a cap, and prospective loss of earnings limited to average net earnings. Periodic payments would be made wherever practicable, subject to review at stated intervals. Those who elect to claim under the no-fault scheme would be disqualified from suing in negligence. The *administration* of the scheme would be accompanied by a separate mechanism for the scrutiny of each claim to ensure that appropriate care had not been transgressed.

The NHS (Compensation) Bill. Eligibility would be based on proof of injury resulting from a 'mishap' in NHS care, and not as a foreseeable and reasonable result of that care or the person's pre-existing condition. *Benefits* would be payable to those who required hospital in-patient treatment for more than 10 days, or who were prevented from engaging in normal activities for more than 28 days, or who suffered significant pain and suffering. The level of benefits is left open for subsequent draft guidance. The scheme would be *administered* by a Medical Injury Compensation Board, and finance would be at a level determined by the Secretary of State for Health.

It should be noted that each of these schemes are restricted in some way to those injured in the course of medical treatment. This means that a disabled individual, to benefit under a scheme, would need to prove that his/her injury was caused in some way by medical intervention. Those who can do so successfully would receive generous compensation (although the degree of generosity varies between schemes) compared with those who cannot, who would need to rely instead on social security benefits. For this reason very similar disabilities can lead to very dissimilar amounts of compensation.

Secondly, no-fault schemes are, by their very nature, imperfect as a means of securing medical accountability. Unlike the tort system, there is no need to identify a particular individual's error in order to qualify for compensation, and so it could be argued that the replacement of the test of fault with the test of causation removes a potentially important source of information about the standard of medical practice. The tort system, for all its problems, is unique in the way it provides patients with a financial incentive to reveal such information. Only if we feel that medical error is somehow unavoidable can we afford to ignore this as a potential disadvantage of no-fault. Recognition of this argument in the NHS (Compensation) Bill was shown by the provision of powers to 'take action' by the proposed Medical Compensation Board, although the criteria by which these powers were to be exercised were left vague.

SOCIAL SECURITY

Under social security, individuals receive support on the basis of the fact and severity of their injury and its consequences, and have to establish neither fault nor cause. Damage suffered following medical intervention is treated in exactly the same way as damage suffered in any other way, including chronic illness and congenital disability, as well as accidental injury. The payment of benefits periodically rather than as a lump-sum allows compensation to adjust in relation to changes in the claimant's circumstances. Moreover, because social security does not require proof of cause or fault, administrative costs are correspondingly lower than with either tort or no-fault alternatives.

In the UK, there is currently a complex range of overlapping benefits and allowances, both contributory and non-contributory, which is available to the disabled and the partially disabled. A small amount of supplementary income-related benefit is payable, based on the claimant's contribution record, but in general the level of compensation is relatively low. The introduction of a comprehensive disability income, comprising a disablement cost allowance together with an income replacement benefit, has been proposed by pressure groups such as the Disablement Income Group (DIG) as a means of reforming the current system of state benefits. Even more far-reaching would be the integration of all forms of disability compensation, both private and public, into a single scheme, along the lines of the New Zealand Accident Compensation Scheme described above, but extended to include illness-related disability as well as accidental injury (Lamb and Percival, 1991). Unlike the New Zealand scheme, this would remove altogether the need to establish causation. The claimed advantages of such a scheme over the fault- and cause-based systems lie in its ease of access for claimants and its relatively low administrative costs. However, because improved accessibility and wider eligibility would increase significantly the number of claims made, the questions of finance and generosity of benefits become critical.

There remain problems with respect to other objectives, such as rehabilitation and prevention of accidents. The payment of reviewable periodic benefits may have an effect on the recipients' recovery, and, where relevant, their return to work. As far as prevention is concerned, the removal of any financial incentive to demonstrate individual or organizational responsibility for damage means that the implementation of any such scheme would need to be associated with an emphasis on alternative arrangements for monitoring standards of care — including those of the medical profession and the health care sector generally.

Conclusion

There is no easy answer to the question of medical injury compensation. There is no doubt that the current mixture of fault-based compensation (through tort) and fact-based compensation (through social security) leaves a large residue of uncompensated losses which fall upon the victim. One possible response to this situation is for victims to bypass the issue of compensation altogether and purchase first-party insurance cover, either explicitly, or as part of an implicit contract with employers. These options have not been reviewed in this chapter, as they might be considered beyond the scope of compensation policy. As we have seen, the latter concerns itself with arrangements to reduce the extent to which patients bear responsibility for their own losses, either by passing these on to doctors, or to the taxpayer.

By characterizing compensation policy in this way, it becomes clear that the exercise is essentially one of redistribution. In principle, the burden of medical injury can be shared throughout the community in a more equitable way without reducing total national wealth at all. The debate over the 'cost' of no-fault alternatives should be interpreted on this view as a debate over the limits to acceptable redistribution. Too high a 'cost' means that society is unwilling to sanction transfers at that level.

However, this is to simplify the issue too much. While the motivation behind compensation policy can be characterized as redistributive, the *constraints* to the effective pursuit of this objective are not simply those relating to social preferences for equity. As we have seen, each alternative mechanism for transferring losses from patients to others may carry with it some penalty in terms of administrative cost, an increased number of accidents due to inadequate prevention, or longer periods of recovery and rehabilitation. These are the 'efficiency' consequences of compensation — the real social costs of each policy. What is more, these efficiency costs may be exacerbated by arrangements for risk-pooling associated with each policy. For example, placing liability for patients' losses on doctors may be intended to strengthen accountability at the same time as improving equity, but the availability of third-party liability insurance will mean that the careful doctor will be forced to pool his liabilities with the careless, unless insurers are in a position to monitor accidents and vary premiums accordingly. It could be argued that the doctor's employer (or his colleagues) are in a far better position to undertake this monitoring process, and that the best way of pooling liabilities is therefore at the organizational level.

Of course, the objectives of organizational accountability and patient compensation could be pursued independently, through hospital regulation and social security, respectively. However, the point that has been emphasized here is that the choice of a particular mechanism for compensating accident victims may imply an 'opportunity cost' in terms of injurer accountability. The information generated by patients pursuing valid claims in relation to organizational failure may be lost if patients have no financial incentive to bring claims. What needs to be considered, then, is the appropriate balance between the (inequitable) 'premium' payable to those who are in a position to produce valuable information for accountability purposes, and the purely compensatory payments relating to the severity of disablement. Determining that balance will be the principal task facing policy makers in this area when the climate for reform improves.

References

Brennan, T. (1992). An empirical analysis of accidents and accident law: the case of American medical malpractice law. *St Louis University Law Journal*. (in press).
British Medical Association (1983). *Report of the working party on no-fault compensation for medical injury*. British Medical Association, London.
Civil Justice Review (1988). *Report of the review body on civil justice*. HMSO, London.
Department of Health (1991). *Arbitration for medical negligence in the NHS*. Department of Health, London.
Fenn, P. and Dingwall, R. (1990). The problem of crown indemnity. In *Health Care UK 1989*. Policy Journals, Newbury, Berkshire.
Ham, C., Dingwall, R., Fenn, P., and Harris, D. (1988). *Medical negligence: compensation and accountability*. King's Fund Institute/Centre for Socio-Legal Studies, London/Oxford.
Lamb, B. and Percival, R. (1992). *Paying for disability: no-fault compensation — Panacea or Pandora's box?*. Spastics Society, London.
Lord Chancellor's Department (1989). *Contingency fees*. HMSO, London.
Royal College of Physicians (1990). *Compensation for adverse consequences of medical intervention*. Royal College of Physicians, London.
Weiler, P. (1991). *Medical malpractice on trial*. Harvard University Press, Cambridge, Massachusetts.

14

Accountability

ARNOLD SIMANOWITZ

The emphasis on the issue of compensation for victims of medical accidents in the past decade by most of those concerned with such accidents has been understandable but unfortunate. Compensation is, of course, important, and where as a result of a serious injury victims have been deprived of the means of supporting and caring for themselves or their dependants, it is of major importance.

For anyone who has been deeply involved in the issue of medical accidents, however, it is the question of accountability which has proved to be of far greater importance to victims. Accountability from the victim's point of view means simply that something is done to ensure that those responsible, if indeed there be any, are required to give an account of themselves, that an explanation is given to the victim or family, and that steps are taken to try to avoid a similar accident happening again.

For the vast majority of victims about whom anything is known, the compensation will not be great and is not of primary importance. A substantial majority of victims in fact never comes to light. In a huge study of medical accidents carried out in New York involving the examination of 50 000 files it was established that for every eight people actually injured by negligence only one brought a claim (Harvard Medical Practice Study, 1990). Action for Victims of Medical Accidents (AVMA) has now dealt with over 12 000 enquiries from people who believed that they had been injured in a medical accident and notwithstanding the media hype that has been given to major awards of damages and the general emphasis on compensation, the majority of those who approach us even today do not initially or primarily seek compensation.

They come to see us in distress, bewilderment, and anger generated by a feeling of helplessness. They want to know what has happened to them and why it has happened; they want to know who was responsible and if someone was responsible then they desperately crave three things. Firstly, they want an apology from that person face to face — not a letter from a faceless bureaucrat apologizing, as often as not, for the fact that patients 'feel' that they have been badly treated. This overriding need for an explanation and an apology is confirmed by anyone who is involved with the problems of patients (see, for example, the report of the Association of Community Health Councils (ACHEW, 1990)). Secondly, they want to know that that person's superiors, employers, or colleagues have enquired into what has happened and, if appropriate, have taken some action — and they must learn what that action is. Thirdly, they want to know that steps have been taken to ensure that what happened does not happen again.

Indeed, it is refreshing to see so often that people in the direst distress who have suffered major injury — very often the loss of a loved one — see as a priority the need to ensure that what happened to them is not allowed to happen again to anyone else.

It is not only because the patient wants an explanation that it is vital that accidents should be investigated. Unless there is a full and independent enquiry when any accident takes place (and the depth and extent of the enquiry will depend on the seriousness of the accident and its likely ramifications but independent remains the key word) the conditions which led to the accident may remain and patients will continue to be at risk. This is and certainly should be as much a concern of the Health Service Administration as of the patient.

A simple example will serve to illustrate how failure to acknowledge an accident or to investigate will perpetuate the conditions which led to it. Over many years a number of women who had undergone Caesarian sections complained that they had been awake during the operation and aware of everything that had gone on and had suffered excruciatingly. Notwithstanding the fact that in at least one hospital there were numerous such complaints, the health carers did not take them seriously and assured the patients that they were imagining what had happened.

Eventually a patient was successful in a legal action for damages (Ackers v. Wigan Health Authority (1991) 2 Med LR 232). She had been able to describe not only the sensations she had felt as the various operative steps were taken, but also the details of the conversations the doctors and nurses were having. Because the subjects she referred to could be verified she was able to satisfy a judge that it had actually happened.

As a result of the publicity that that case attracted a large number of claims were made. The health professionals were not only forced to acknowledge that such problems could and did take place but they examined their procedures and identified exactly what was causing the problem. They were then able to alter their procedures and training accordingly. The result is that this particular problem has virtually disappeared. I believe that it is unlikely that anyone would want to argue that it should be legal action which should bring about an improvement in treatment of this kind, but in this instance legal action did lead to important changes in clinical practice.

It is all that is implied by these concerns that is encompassed by the term *accountability* in the sense in which I shall be dealing with it in this chapter. There are people who take the view that a professional needs to be accountable only to his or herself. Indeed when Norman Tebbit MP received a letter from one of his constituents urging his support for an independent enquiry into medical accidents he replied: 'If, as your letter states, "little" is said about how to avoid such accidents in the future as well as the accountability of doctors, I ask said by whom? I cannot imagine there is a doctor or surgeon in the country who is not conscious of the need to both avoid accidents and to be accountable to his or her patients.' (personal communication).

There are two different ways of looking at what is meant by accountability. The first is by way of academic definition and for this I can do no better than rely on the words of Professor Margaret Stacey, Emeritus Professor of Sociology at the University of Warwick: 'There are so many ways in which a doctor may be held to account for her/his actions; for clinical actions to individual patients and, in medical audit, to colleagues; at law, in terms of obligations to patient or employer; to the profession for his/her behaviour; to employers for the money spent and the priorities adopted in treatments. Furthermore, as a collectivity the profession is held to account to the public at large for the quality of medical care in general and particular' (Stacey, 1989).

The second way to understand accountability is by describing the experiences of patients during treatment. This can be achieved best by way of the following anecdote. A young mother-to-be was threatening miscarriage at 24 weeks and was admitted to the maternity ward for observation. One morning after she had been there for a week the consultant arrived followed by his retinue and they gathered round her bed. They proceeded to discuss the 'case'. When the woman tried to ask questions she was ignored. At the end of the discussion the Consultant left followed by his admiring retinue. Not a word had been addressed to the young woman.

Now of course not all doctors are like that but I am bound to say that having related this story — which took place in 1990, not in 1950 — to numerous doctors they all confirmed that this was by no means exceptional behaviour. Somehow that behaviour has to be corrected by some person or body that is independent of the profession itself.

There is no system at present in existence that can ensure any measure of accountability over behaviour of that kind and certainly the General Medical Council, which is supposed to regulate doctors, has no plans to address it.

That kind of behaviour may not seem serious as, on the face of, it does not involve standards of treatment. Nevertheless, it is behaviour that can lead to deficient treatment simply because so much of health care depends on communication and the relationship between patient and doctor.

It is against that background that we must look at how the systems for compensation but more importantly for accountability work, both in the UK and in other jurisdictions. It is particularly important to look at the USA where until recently litigation has been the chosen method for enforcing some form of accountability, and at New Zealand and the Scandinavian countries where the system of so-called no-fault compensation has impacted on accountability to a greater or lesser extent.

Accountability in the UK

In the UK the problems for patients who seek to answer the questions I have posed above are not quite insurmountable but they are sufficiently daunting, frustrating and unsatisfactory as to make the whole process a mockery in all but a small percentage of the cases.

The problems relate to two aspects — accessibility and effectiveness. Insofar as accessibility is concerned the main obstacle to a patient actually getting the problem dealt with is the proliferation and complexity of the systems of enquiry. This problem has been dealt with at length elsewhere (ACHCEW, 1990) and needs only to be briefly stated here. If patients in the NHS wish to complain, to find out what happened, or to try to obtain an apology, they have a choice of six systems depending on the exact nature of the complaint or where the treatment took place — the NHS hospital complaints procedure; a complaint to the Family Health Service Authority (FHSA); a complaint to the Health Services Commissioner (Ombudsman); a complaint to the General Medical Council; and a complaint to the United Kingdom Central Council for Nursing Midwifery and Health Visiting (UKCC). Finally, if the patient wishes, instead of or in addition to simply complaining, to seek compensation, then they must proceed by way of litigation in the Civil Courts.

The procedures for complaining about the behaviour of a medical practitioner are complex and confusing to the patient, and beset by unrealistic time limits and unnecessary formality. As regards the only system which has any real teeth insofar as accountability is concerned, namely the General Medical Council procedure, which is the only one which includes the potential sanction of removing the right of the practitioner to practise, the patient begins with the hurdle of having to swear an archaic Statutory Declaration and the complaint is then filtered by various arcane and secretive systems. Because of this, and because of the very restrictive meaning given to the term serious professional misconduct, only a tiny proportion of complaints ever reaches a hearing (Robinson, 1991).

As regards a complaint to the FHSA this must be brought within thirteen weeks of the incident complained of and no provision is made, save a discretionary one at the whim of the Chairperson of the Service Committee, for the fact that in many cases because of the lack of openness on the part of the medical profession patients won't even know about the incident until long after it takes place: for example, a particular drug may have a bad reaction and the patient may not know for a long time afterwards that it should not have been prescribed.

Effectiveness of the procedures

It is the experience of any person or body dealing with complaints of patients that all the systems are highly unsatisfactory from the patient's point of view. Although there may be many reasons why the number of complaints in the various systems is minuscule in relation to the actual number of 'treatments' that take place (see the Harvard Medical Practice Study, 1990), I would suggest that one of the major reasons is the belief among patients that they will get little if any satisfaction out of complaining.

THE HOSPITAL PROCEDURE UNDER HM(88)37

Insofar as AVMA is concerned, if there is any chance that a legal investigation can be mounted the advice to the patient will be to use that route whereby at the very least a truly independent assessment from a consultant instructed by the patient's lawyer will be obtained.

The main reason for this is the secretive nature of the hospital investigation. At no time is the patient a party to that investigation. Since the coming into effect of the Access to Records Act 1990 patients can at least obtain their records. However, what the patient will still need and invariably will not get, unless the independent consultants undertake it, is a doctor to go through the notes with them explaining the implications of the entries about what has happened to them. He or she is expected to take on trust that the method of enquiry has been adequate and unbiased as well as the decision of the 'independent' consultants. Given that the complaint has been made in circumstances which must place the patient's trust in medical practitioners under suspicion this may be a little too much to expect of the patient.

The patient is not privy to what investigations are made, who has been interviewed or what they have said; does not have an opportunity to challenge any statements; and does not have access to the notes which the consultants have read. They are not even entitled to a full report from the consultants themselves on their findings but only a bowdlerized version from the hospital administration.

THE SERVICE COMMITTEES

The biggest drawback to the effectiveness of this procedure for the patient is the nature of that procedure itself. It is not meant to be an enquiry into behaviour towards a patient. It is simply a matter of employment contract and whether there has been breach of that contract.

Whilst the patient may have triggered the complaint, the issue is between the employer (the FHSA) and the employee (the doctor). This is never explained to the patient who remains throughout under the impression that it is their complaint which is being dealt with. This greatly circumscribes the enquiry. Additional problems for the patient are caused by the fact that whilst they cannot afford to be professionally advised, the doctor invariably has the assistance of the relevant defence organization.

THE GENERAL MEDICAL COUNCIL

It is the belief of most patients that a complaint to the GMC is a waste of time. Apart from the problems of actually getting the complaint properly investigated, the very legalistic manner in which the proceedings are conducted and the burden of proof discharged leads patients to believe that the procedure is concerned far more with the rights of the medical practitioner than with those of the patient. This belief is re-enforced by the discussions initiated by the GMC to introduce a performance review ostensibly to deal with issues which may fall short of the definition of serious misconduct.

Objections raised by the Association of Community Health Councils and by AVMA suggest that in patients' eyes the new procedure will, in many instances, lead to cases which at present at least have a public hearing being dealt with under a new procedure which does not result in a hearing. The patient will in any event once again be excluded from the decision making and will not even know what steps, if any, will be taken to ensure that the behaviour complained of will not be repeated.

THE OMBUDSMAN

The Ombudsman procedure is more satisfactory both from the procedural point of view and in terms of its effectiveness. That is not to say that the position would be improved if the Ombudsman's role were to be extended to deal with matters of clinical judgement and/or professional misconduct as some are urging.

The nature of the profession, and indeed any profession, is such that the procedures which are applicable to administrative matters would not be appropriate to the issues of the accountability of the health professionals themselves.

LEGAL ACTION

It is often argued, by those who would ensure that the tort system is not abolished, or even watered down in favour of some form of no-fault system for compensation, that legal action does not only ensure compensation in cases of negligence but also exercises a form of accountability. This is true only to a very limited extent. Cases where there has been clear negligence rarely if ever go to court. They are settled without a hearing and if any steps are taken to ensure accountability they will be taken outside the legal proceedings.

Cases where the patient's legal adviser cannot establish a cause of action on a legal basis will not be investigated at all. The only cases which actually have the benefit of a full investigation in Court are those where legal liability is hotly contested and on occasion it is the 'innocent' doctor who is subjected to this inquisition.

DO OTHER COUNTRIES ORGANIZE THINGS IN A BETTER WAY?

Most countries which have tackled the issue of medical accidents have looked only at the compensation aspects. It is not appropriate in a chapter on accountability to look too closely at the various no-fault compensation schemes. However, it is important to see what impact those schemes have had on the question of accountability and whether they have specifically addressed that issue in any way.

The leader in the field, and one to which those in Britain who favour a no-fault system always look as an example, is Sweden. Sweden introduced their scheme in 1975. What must be borne in mind when considering the Swedish scheme is that it was never intended as a comprehensive scheme for compensating victims of medical accidents. Sweden has for years had an advanced system of welfare and the no-fault scheme was seen simply as a 'top-up' to that system for victims of medical accidents.

What makes Sweden's scheme unique is that the Swedish appear to have recognized, long before any other country was even aware of the issue of medical accidents, that patients would want more than simply money. Accordingly, at the same time as they set up the compensation scheme they also set up a system of accountability known as the Medical Responsibility Board.

There are major defects in both the Swedish compensation system and the accountability scheme. Insofar as the former is concerned the issue of causation, which would loom large in a system which was the main or only system for compensating victims of accidents, is dealt with in a somewhat hit and miss fashion. Furthermore, because the compensation is no more than a 'top-up' to welfare payments, the Swedish are not too concerned with defining fault closely and decide whether claims should come within the scheme in a very *laissez faire* manner. Notwithstanding this it is not generally appreciated that 60 per cent of claims in Sweden are rejected (Brahams, 1987). Insofar as the accountability side of the Swedish scheme is concerned there is considerable scope for bad practice to be overlooked because of the failure of any real challenge on behalf of patients to the findings made by the doctors.

Probably the main defect in both the compensation and accountability aspects is one which is seen by some commentators as their main advantage —

namely the fact that there is no connection between the two. The theory
behind this separation is that doctors are to be encouraged to report their own
mistakes to the compensation body in order to ensure that patients receive
their compensation and that they would be reluctant to do so if they felt that
would expose them to the jeopardy of the Medical Responsibility Board.

This theory is subject to two criticisms. Firstly, it is my view that this
separation may bring hardship to patients and is not in the interests of the
Health Service. One can envisage the unacceptable situation where compensation is paid after the compensation board receives information disclosing
serious defects in practice, which make a doctor a danger to the public.
However, the Medical Responsibility Board may fail to discipline the doctor
in any way because of lack of evidence.

Secondly, the theory exposes an attitude towards the medical profession
which is unhealthy for society and which has to be changed if patients are to
be seen as partners of health professionals rather than simply as recipients of
their skills. It is not acceptable that health carers should be able to say that
they will not admit to mistakes which could actually help their patients unless
they are given immunity from disciplinary enquiry.

What is needed is a change of culture whereby health professionals accept
that they are accountable. Keeping the two systems separate will simply
perpetuate the idea that they are not accountable.

The problem in trying to use the Swedish system (and indeed the Finnish
and Norwegian systems which are based on the Swedish one) as a precedent
for Britain, or any other country, is twofold. In the first place, as Donald
Harris, director of the Socio-Legal Centre at Wolfson College, Oxford, said
when summing up the deliberations of a two-day conference on systems of
compensation for medical accidents (Harris, 1992), the Scandinavian systems
are suited to their culture and would not necessarily work for other countries.
Secondly, the Swedish system was set up in advance of any demand by
patients in that country. The Swedish people, because they had not developed
a patients' movement which understood the issues thoroughly as has
happened in Britain, have not been able to mount the challenge to the
entrenched views of the medical profession which would still be prevalent in
the doctors deputed to investigate the issues by the Medical Responsibility
Board.

This situation, where inadequate solutions are presented for the problem of
medical accidents in advance or in the absence of informed demand by, or
indeed as a result of failure to heed the views of, the people actually affected
by the issues is reflected in the experience of other countries as well.

The New Zealand system of no-fault compensation was introduced in
1972 as a result of the Woodhouse Commission report, which not only
looked solely at compensation issues to the exclusion of accountability, but
was concerned with accidents and injuries generally. No attention whatsoever

was paid to the special issues thrown up by medical accidents. The major problem that was, as a result, ignored is the fact that in no other case is it the very profession that has caused the harm that is called upon to adjudicate on the nature and provenance of the injury caused.

There were two effects of this. Firstly, not only were many claims for compensation wrongly rejected because of the evidence of doctors (see, for example, the comments of Margaret Vennell, Senior Lecturer in Law, University of Auckland and a member of the Board of the Accident Compensation Commission (Vennell, 1990)). Secondly, there was a lot of evidence to indicate that the attitude of many doctors became somewhat cavalier in the knowledge that they were no longer exposed to legal action. Indeed, the medical establishment in New Zealand itself became concerned about that problem (Vennell, 1990). The concern of the medical establishment led to a number of changes in the New Zealand system of accountability, but it still does not come anywhere near the type of full accountability which would satisfy the public that the medical profession is really accountable for its behaviour.

To be fair to both the medical and legal professions it must be recognized that it is not only these professions which have tended to see the problem of medical accidents in terms of compensation alone. Throughout Europe, where patients have come together to try to do something about the problem, they have themselves perceived it as primarily one of compensation. For example, in France victims of medical accidents who have become frustrated by the difficulties in securing the compensation to which they are convinced they are entitled have for some years now been trying to change the system of compensation without looking at accountability at all. There is an organization fighting in this area called Association De Secours Et De Sauvegarde Des Victimes Des Actes Medicaux (AVIAM), which has been seeking the introduction of legislation to create a form of no-fault compensation for some time but without success.

It is not surprising that initially the problem is seen by many victims as being solely about compensation when it is brought home forcibly to them that they are, unlike victims of any other accident, unable to secure compensation. It takes some time for them to realize that the reason for their failure to secure compensation is the very lack of accountability of the medical profession which enables it to obscure the causes of an accident, to refuse to answer questions, and to stick together in such a way that nobody can find out what really happened. It is only later, too, that they also realize the effect that these closed ranks can have on standards (see the discussion in connection with the British position below).

In Germany there is a much clearer perception of the real issues. The Allgemeiner Patientenverband has been working in the field since the late 1970s. Their main thrust has been to help individual patients to secure compensation but, working in the field for so long, they could not avoid becoming aware that the total lack of accountability underlies the failure to provide compensation to victims of medical accidents. It is likely that the fact that the activists in that organization are members of the medical profession has assisted it in identifying the underlying problems.

Probably the country which has chosen the best way of tackling the problem in the absence of a vibrant patients' body on which it could rely for information is Denmark. The Danes did not start from the angle of compensation. What they tried to institute was a form of inspectorate of the health profession in order to maintain standards. I shall argue later in this chapter, although it really should not require any argument, that the only way of tackling the problem of medical accidents effectively is to tackle the issue of standards head on. Perhaps the Danes have not done exactly that but the system of Health Commissioners (medical practitioners working full time for the state) who have the right to investigate not only accidents but health practices is clearly far preferable to simply looking at the question of compensation.

As far as compensation is concerned, Denmark has recently introduced a system of no-fault compensation based on the Swedish scheme and it remains to be seen how that will impact on the existing system of inspection described above and whether it will in fact adversely affect the limited accountability established by that system. However, the position in Denmark for victims of medical accidents is still sufficiently unsatisfactory for it to cause major concern for at least one of the Commissioners and to lead to the Danish being in the forefront of the campaign to secure a united front on the issue in Europe, which I shall next describe.

In February 1992 at a meeting in London, representatives from a number of countries concerned about the problems of victims of medical accidents met to consider what action could be taken on a European front to deal with those problems. It was agreed to set up a European organization, the European Society for Victims of Medical Accidents. The problems of victims were identified as being virtually identical in all the countries represented, namely Britain, France, Germany, Denmark, and Sweden, as well as those which were unable to attend but had sent apologies and given an indication of their concerns, namely Austria, the Netherlands, and Switzerland. These problems were primarily accountability and then compensation. It was the power of the health care professions in relation to patients which was the stumbling block for patients who had suffered injury or who had complaints about their treatment.

Britain's position among all these countries is a unique one. It alone has had for many years two effective organizations which have been concerned with the issues of complaints and compensation. Action for victims of Medical Accidents (AVMA) was established in 1982 and since that time has had direct or indirect contact with some 25 000 people who have been or believed themselves to have been victims of a medical accident. The Association of Community Health Councils has, through its constituent members the Community Health Councils of which there is one in every Health District, had experience of many more people who have complained about the health care they have received. These two bodies have identified the reforms that are required to ensure that health care professionals can be held accountable to the people for whom they care.

Their proposals for such reform represent, in my view, the only basis on which the radical re-appraisal of all the systems can take place to bring about not only justice for those patients who have a grievance but also the control over standards which is so vital to ensure that the number of accidents and complaints are reduced.

A health standards inspectorate

The AVMA/ACHCEW proposals involve the creation of a statutory body, A Health Standards Inspectorate. This Inspectorate would be the sole point of entry for any complaint from a patient of whatever nature. No longer would the patient have to decide into what category the complaint fell in order to identify where to go to complain. Nor would there be complicated legalistic procedures such as the Statutory Declaration demanded by the General Medical Council.

The Inspectorate would be controlled by the policy-making Board which would comprise a majority of lay members together with health care professionals of various disciplines as well as representatives from other appropriate bodies.

The main thrust of the inspectorate would be to improve and maintain standards. The inspectors would therefore have rights of access to all sites where health care is delivered, whether public or private, and it would be an integral part of their work to inspect these sites and take action in respect of any deficiencies found.

As part of the inspectorate there would be a complaints body to which complaints could be made by patients at a local or national level or brought by one of the inspectors. This body would establish panels which would hear complaints that could not be resolved without a hearing.

There would also be a body which would deal with compensation issues. The question of what injuries would be compensated has caused problems of definition in the past because, I believe, compensation has been the only aspect of a medical accident which has been amenable to proper investigation. It is my firm belief that once the issue of accountability has been properly addressed, the public at large will be prepared to accept that there will be a category of accident which will not be compensated. It is in those circumstances with the general needs of complainants being met by the comprehensive approach of the Inspectorate that the compensation body will be able to adopt as the basis for compensatable accidents 'an injury suffered by a patient as a result of a mishap in care and not as a foreseeable and reasonable result of that care or the person's pre-existing condition'. This is the definition adopted by Clause 2(5) of the National Health Service (Compensation) Bill introduced by Rosie Barnes MP in 1991.

The compensation proposals do not meet the objection that in terms of need, defining the right to compensation in terms of how the accident happens is unfair. There is, in my view, an incontrovertible argument for saying that all disabled people should have an adequate measure of support to enable them to have the maximum quality of life within their disability. If that position were to be achieved insofar as victims of medical accidents are concerned without an adequate system of accountability, I have no doubt that would leave the public dissatisfied because only one aspect of the problem would be addressed. The attractiveness of the AVMA/ACHCEW proposals is that if and when Society reaches the view that it can afford to support disabled people adequately then the Health Standards Inspectorate could be adapted to this situation without the need to demolish the whole system. Similarly, if there comes a time when it is believed that accidents other than medical accidents should be compensated in this way the system could easily be adapted to accommodate this.

The Inspectorate will have to be comprehensive in the interests of both simplicity and the free flow of information. For this reason there would have to be a body to deal with complaints about administrative matters, and one to deal with disciplinary matters, within the inspectorate. These bodies would operate in similar ways to the way in which the Ombudsman and the General Medical Council operate at present, but they would be subject to the lay representation and overall control of the Board as the other bodies would be.

Proposals of this kind are bound to generate controversy and debate, particularly in regard to matters of detail. I would suggest, however, that the principles involved — a universal system to deal with all complaints subject to parliamentary control and with a majority of lay members — is the way forward to achieve the kind of accountability for which citizens in the last years of the twentieth century are looking.

References

ACHCEW (1990). *NHS complaints procedures — a review*. Association of Community Health Councils in England and Wales.

Brahams, D. (1987). No Fault Compensation based on patient insurance. *Lancet* (21 March), 698.

Harvard Medical Practice Study (1990). *Patients, doctors and lawyers: medical injury, malpractice litigation and patient compensation in New York.* Report of the Harvard Medical Practice Study. HMPS, Cambridge, Massachusetts.

Robinson, J. (1991). *A patients voice at the GMC.* Health Rights Ltd., London.

Stacey, M. (1989). *Medical accountability.* Background paper prepared for a meeting organised by the King's fund Institute and the Centre for Medical Law and Ethics: King's College, London.

Vennell, M. (1990). *Medical injury compensation under the New Zealand Accident Compensation Scheme and medical responsibility.* Paper presented at the International Comparative Law Congress, Montreal.

15

Safety in medicine

CHARLES VINCENT, ROBERT J. AUDLEY,
and MAEVE ENNIS

The range of adverse outcomes considered in this book is far wider than is often associated with the term medical accidents. Most authors, including ourselves, have widened the definition to examine a wide variety of examples of substandard or suboptimal care. Not all these adverse outcomes are the result of error, and fewer still are the result of negligence. Even where an individual error can be highlighted, it is usually only one part of the complex web of causes and conditions that underlie many accidents. For this reason, a routine audit of a particular procedure, valuable though it is, does not have the same power to unravel the complex of factors involved in many accidents as does a systematic analysis of their multiple causes.

Although we believe that this book demonstrates the important contribution to health care that comes from the study of accidents, we are very aware of its limitations. A particular concern is that only three medical specialties have been examined in detail, each of which has, probably unfairly, received much attention as a source of accidental injury to patients. These specialties have been at the forefront in the analysis of errors and accidents and of attempts to monitor and improve safety. The choice of specialties inevitably introduces some bias; each has its own associated risks and its own special problems. Delayed or incorrect diagnoses, for example, may be far more common in general medicine or general practice than in obstetrics, surgery, or anaesthetics.

We do not propose to summarize the contents and conclusions of the preceding chapters because the authors have already given concise accounts of their particular concerns. There are, however, a number of recurring themes which seem to us to be especially important. In some cases there are unexpected correspondences and alliances between individuals writing independently from very different perspectives. This final chapter will briefly state and discuss these overarching themes. We have not often referred to specific authors by name, but it will be apparent that many of the ideas and proposals derive from their contributions.

A shift of emphasis from the individual to the organization

As the first chapter demonstrated, the investigation of accidents in other fields has often revealed widespread deficiencies at an organizational level. In this book the observations of both clinicians and psychologists suggest that this is equally true in medicine. From this perspective doctors, and others directly involved in patient care, are often the inheritors of latent failures — faulty policies, unfortunate management decisions, poor communication, previous clinical errors, and the like. This is not to say that individuals do not make mistakes, or should not take responsibility for those mistakes. However, while it may be convenient to hold one individual responsible for an accident, he or she may in reality be little more than a scapegoat. A full analysis may show that they may have been placed in a situation where errors were, to some degree, likely to occur. Accident-proneness or a characteristic careless-ness, while serious, is probably not of major concern on the majority of oc-casions: mistakes and errors tend to be made in particular circumstances rather than by particular individuals.

Although it is important that individuals accept responsibility for error it is often both unjust and pointless to blame those individuals, and it lessens the chances of any constructive response. Blame, if applicable, should generally be distributed more widely and not be solely placed on the unfortunate person whose action was the final link in a causal chain. Blame, however, is certainly appropriate in the case of those actions that Reason terms 'viola-tions' — deliberate or irresponsible departures from accepted procedures. Acting beyond one's competence, failing to call more experienced colleagues, failing to take account of the views of other professions, and failure to carry out checking procedures have all been noted at a clinical level. Imposing unacceptable staffing levels might be regarded as a managerial equivalent.

The tort system tends unfortunately to mean that the actions of a parti-cular individual are singled out for scrutiny while the actions of senior staff, hospital management, or government are seldom called into question at any hearing. The move to Crown Indemnity, where liability for injury to patients is carried by the Health Service and hence by government, is welcome in that it removes some of the emphasis on the role of individual staff. More im-portantly it gives organizations, whether individual hospitals or larger bodies, a direct incentive to promote safety. Tort is, as several people have argued, at best an inefficient and expensive way of ensuring accountability quite apart from the effect it has on individual clinicians and the practice of medicine. With the change to organizational liability tort may make marginally more sense, which is not to say that further reform is not necessary.

The principal areas of concern

Many of the error-producing conditions identified in the analysis of industrial and transport accidents have also been independently identified by clinicians as major areas of concern. This correspondence suggests that the methods of analysis that have proved so fruitful in other areas may be equally valuable in medicine. The same broad categories emerge: high workload, inadequate supervision or training, poor interface design, a stressful environment, and mental states such as fatigue, boredom, and depression. Staff characteristics and selection have also been considered although McManus has argued that selection, while important, has potentially less impact on safety than effective training.

All the clinical authors are agreed that training is a key area; it would appear that both training and supervision are inadequate. There has long been an attitude in medicine that training happens naturally during clinical experience, and that formal instruction, particularly after qualifying, is generally unnecessary. Combine trial-and-error learning with a lack of consultant supervision and the risks become apparent. Add work overload, lack of sleep, lack of awareness of hazardous situations, a degree of over-confidence stemming perhaps from a desire to prove oneself, and the potential for serious accidents becomes apparent. These factors are only partially in the control of individual doctors; many also require action at an organizational level, both from senior clinicians and management and from government.

Increasing safety

Most authors have stressed that medical accidents are little researched and consequently little understood, although there are strong suspicions about some of their likely causes. If a greater understanding is to be reached, the systematic analysis of errors and accidents will have to acquire a higher priority with the Research Councils and Charities who fund medical research and also with the Department of Health. It would be premature, in these circumstances, to offer detailed proposals for increasing safety. However, several important suggestions have been made, some by several authors independently.

Risk management in the National Health Service is a very recent innovation and is, as yet, primarily a medico-legal matter. Staff are encouraged to report incidents that may lead to litigation so that the necessary statements can be gathered and information collected before those involved have moved away or forgotten the incident. However, if this information is only used in a medico-legal context it may have no impact on clinical practice. Risk management at the moment is primarily a matter of damage limitation, although any moves that lead to a more rapid processing of claims are obviously welcome, provided they do not aim to simply thwart justified claims. If risk management is to extend to actually increasing safety, then it must be directly linked to clinical practice. The adoption of a system of confidential reporting of dangerous errors or conditions, whether or not likely to lead to litigation, would be a valuable innovation in individual hospitals and departments. The problems with such systems are similar to those of critical incident studies — maintaining interest and motivation. However, if the results were taken seriously and acted upon, this might be overcome. Doctors, and other clinical staff, might then recognize the importance of such systems for communication with colleagues. Management would need to give time and resources to such safety measures, even though they might not directly lead to a quicker throughput of patients.

A step beyond confidential reporting would be safety audits, already developed and used extensively in many branches of industry. Although the investigation of errors and accidents is valuable, safety is not just a matter of reacting to the latest disaster, and a continual safety programme is needed. Accidents are, by their very nature, not directly controllable and it is crucial to distinguish those accident conditions that are potentially modifiable from those that are not. It is not yet clear what form safety audits of health care might take. The defence societies are now offering brief inspections of departments and specific procedures followed by a report on potentially dangerous situations. Although outside experts (doctors, nurses, human factor specialists, and lay people) are important it seems critical that a major organization such as the National Health Service should set up its own monitoring procedures. Many of these approaches, including confidential reporting, are already in operation in respect of drugs and the safety of medicines. The same principles do not, however, seem to have been extended to the actual delivery of health care. As Wilson points out it, is curious that the Health and Safety Inspectorate has powers to demand certain standards of safety for staff, but its protection does not extend to patients in respect of, for instance, deficient equipment.

A comprehensive safety programme must cover all aspects of health care and, particularly importantly, all professions. Disciplinary measures within professions, such as the General Medical Council's recent proposals for examining clinical competence, do not meet this requirement and in any case are primarily disciplinary in nature. As we have seen, serious violations of professional standards are important in only a minority of medical accidents. The Royal Colleges do inspect training facilities but that is a quite limited remit and, as Wilson has pointed out, their only sanction (withdrawal of approval for training) is a rather drastic and cumbersome method of influence. The Health Standards Inspectorate proposed by Simanowitz certainly fulfils some of the functions of safety programmes, though its basic aims are slightly different. However, it would respond to complaints and would attempt to influence clinical practice. Safety programmes should surely, although using outside expertise, also be organized within each hospital or district and be primarily proactive rather than simply reactive. Safety programmes would identify areas of weakness, intervene, and monitor improvements; the same aims as conventional medical audit, but with a different focus. The separation of purchasers and providers means that purchasers of health care could request that providers provide evidence that safety is being actively monitored.

Specific measures

In some cases the analysis of accidents points to immediate deficiencies. In-adequate training in key skills has been highlighted throughout this book, and a number of authorities have called for more direct consultant involve-ment and supervision of junior staff. We suspect that many senior doctors, and probably hospital management, are unaware of the extent to which junior doctors are called upon to act beyond their competence simply because there is no one else available. Clearly these issues cannot be resolved entirely within the profession. An increase in consultant posts, for instance, will need government action. Linked to the need for better training and supervision is the question of error-producing working conditions, particularly the excess-ive hours that junior doctors are required to work. It is a curious paradox that it illegal to drive a coach-load of healthy passengers without regular rest, yet it is considered acceptable to care for a ward full of desperately sick patients when close to exhaustion.

Another important theme that emerges throughout the book, although in different forms, is that many aspects of medical practice should make much more use of aids to memory and clinical judgement. In all three clinical chap-ters suggestions are made for the increased use of *aides memoire*, guidelines, and formal protocols. Scurr even suggests that surgeons might be restricted in the kinds of operation they are allowed to perform. Protocols have been resisted because they are perceived as limiting clinical freedom, exposing staff to litigation, or restricting innovation. However, as Drife points out, protocols can allow clinical freedom while still setting the limits on safe practice. The case for protocols, already used in some circumstances (management of head injury, for example) seems strong. As Wilson points out, they may be especially important in emergencies when there may simply not be time to think through an unusual situation. An experienced doctor can react quickly, while a junior may flounder in the absence of an already rehearsed and formalized emergency drill.

Clinical decision analysis takes this argument a step further to suggest that both guidelines and specific clinical decisions should be derived, not simply from a consensus about the best course of action, but from a formal analysis of the associated costs and probabilities of different courses of action. Clini-cal decisions are broken down into specific cognitive tasks, each of which is made separate and explicit. McManus argues that doctors should be trained in the process of decision-making as an essential clinical skill. Dowie suggests that such an approach will not only improve clinical decisions but will alter, or even eradicate, the very concept of a medical accident. The related devel-opment and use of computer-based expert systems to aid clinical judgement and decisions must also be given greater consideration for complex problems.

What began as an analysis of accidents has led to suggestions from several authors of a general change in the way medicine is practised. More guidelines and less scope for individual clinical freedom may appear less exciting and even less rewarding, but it would surely provide greater safety for patients.

Communication and the need for openness

An ability to communicate and to listen is an essential aspect of clinical skills and it is gradually becoming accepted that specific training in communication skills is essential. In spite of this the role of poor communication, both between professionals and between patients and doctors, has emerged in almost every chapter. Clinicians have drawn attention to the problems that arise when patients have unrealistic expectations or are not fully aware of the risks associated with particular procedures. Patients need to understand what medicine can and cannot achieve; unrealistic expectations have been held up as a reason for the increase in litigation. Whatever the truth of this specific claim it does appear that many patients do not appreciate the inevitable risks of many medical procedures. The responsibility for understanding being reached lies partly with the doctor and partly with the patient. It is not at all clear how much information it is useful for a doctor to provide before an operation, for example. A greater openness about the risks of medicine and the ordinary human fallibility of its practitioners might be beneficial in that it would encourage patients to evaluate treatment choices more carefully. Very sick people naturally and importantly entrust themselves to their doctors and necessarily rely on their decisions. For someone able to consider the possible benefits of an operation or a course of medication it might be wise not to devolve all responsibility onto the doctor and to adopt a more critical, evaluative attitude to the treatment on offer. This has the potential of course to make some patients more anxious which, it might be argued, could have deleterious effects on the course of treatment and recovery. However, allowing an appropriate level of anxiety in the face of a major decision about an operation may be preferable to a naïve faith and optimism.

Communication assumes a special importance when things have gone wrong. Patients may blame doctors not so much for the original mistakes, as for a lack of openness or willingness to explain them after the event. A valuable feature of the tort system, seldom remarked on but much appreciated by patients, is that if they embark on litigation and successfully proceed beyond the initial stages, their case is reviewed by a truly independent expert instructed by their solicitor. Unlike the complaints procedures the patients sees the clinical report, has access to the notes, and may be able to discuss their case with the expert. In the last decade it has become acceptable to give such opinions on the patient's behalf, and this can potentially be of great benefit to them. Many patients, after reading such reports, experience a feeling of relief that their suspicions were confirmed and that they were not wrong to persist in their complaints. Equally, where they have misunderstood some aspects of the clinical process, an independent expert may be able to clarify matters for them. Either way, it is unfortunate that such independent experts cannot be brought in at an earlier stage, perhaps without incurring the emotional and financial costs (for both sides) associated with litigation.

Patients' influence on health care

Many injured patients express a wish to ensure that what happened to them does not happen to anyone else. Fenn has pointed out that tort, whatever its faults, does give patients an incentive and a means to provide information on substandard care. Vague assurances that action will be taken, their remarks have been noted, and so forth, are unlikely to satisfy anyone with a justified complaint. They naturally wish their views to be taken seriously. This is understandable and accords with the current emphasis on the rights of the consumer in relation to government and other large organizations. What is also apparent from several chapters, and less accepted, is that patients may have something useful to contribute as well as simply having a right to be heard. Just as clinical reviews of accidents provide valuable insights into quality of care, so do the accounts of injured patients who, at the moment, are too easily dismissed as angry, unbalanced, and litigious. A great deal of useful information about the quality of health care is lost because procedures for making complaints are so difficult for patients to comprehend. Orr's chapter clearly sets out the different procedures and mechanisms but shows that, while staff dealing with complaints may be well-intentioned, it may be difficult for the patient to navigate through the system. Complaints procedures, in spite of attempts to make them more accessible for the patients, are still slow, complex, and often frustrating and bewildering. They seldom lead to any real assurance that changes to clinical practice have been made, whether this concerns the re-training of individual members of staff or changes in procedures or lines of communication. The assurance of changes in practice after justifiable complaints is of course one of the primary aims of the Health Standards Inspectorate.

An effort to actually ask patients about specific deficiencies or inefficiencies in care, rather than just general satisfaction, could provide important information on common errors and areas of particular concern. Whatever organization is concerned with safety and accountability needs to ensure that the observations of patients are taken seriously as providing useful information about standards of care, and even some aspects of clinical practice, rather than simply being part of an exercise in customer relations.

The cost of accidents

Great attention is given to the immediate costs of medical accidents, particularly the spectacular awards for babies injured at birth, and to the costs of litigation. Fenn has estimated the budget for the National Health Service for 1991/92 for litigation is in the region of £60 million, dwarfed by the £5 billion spent on disability benefits but substantial nevertheless. There are, however, many other costs associated with medical accidents, both human and financial.

Many patients suffer increased pain and disability after errors in their treatment, generally for only short periods but sometimes for life. Their relatives too may suffer immensely after an accidental injury or death. Patients may initially experience symptoms similar to those seen after other accidents and traumas, but in the long term are often depressed, angry, and bitter. Many patients will accept that accidents do happen, and can sometimes accept the initial error. Often their anger is concerned more with their treatment after their injury; with the lack of explanation or with simply not being believed. Some are consumed by anger, and it is possible that this degree of feeling impedes recovery and is damaging in itself.

Doctors, especially young doctors, may experience shame, guilt, and depression after making a mistake. Feeling responsible for injuring a patient appears to be one of the main sources of stress for overworked juniors. A striking feature of the book is how often the word blame appears, in very different contexts. Blame, as we observed earlier, may be inappropriate, in that doctors may be the inheritors of others' failures. Doctors may feel angered and distressed by being blamed. Litigation can feel like an assault — interestingly a word that some injured patients use to describe their sense of outrage.

These feelings reveal the deeply personal nature of the relationship between doctor and patient, which would be acknowledged by many people. Less easy to acknowledge and to understand is the other side of the coin — the disruption of that relationship after an injury, which can be experienced by either side as a serious betrayal of trust. Words like betrayal and assault may seem melodramatic, but are for many people an exact expression of their feelings. It is often not appreciated how deeply people may be hurt in such circumstances—whether patients, doctors, or anyone with a responsibility for caring for others.

The consequences of accidents are, then, far greater in human terms than is generally realized. However, the financial costs are also vastly greater than the immediate costs of litigation and compensation, especially as only a small proportion of injured patients ever receive compensation. Any serious injury leads to greater costs for the family involved and for the state in terms of disability benefits and greater reliance on health services. Injury to patients means increased use of hospital resources and additional time with staff. Such patients may become more difficult to help if they have become angry or mistrustful of the medical profession. A doctor whose confidence has been impaired by making an error, or believing that they are responsible for the injury or death of a patient, will suffer personally which in turn may affect their subsequent clinical judgement and performance. At the worst they may abandon medicine as a career. Firth-Cozens argues that the provision of some confidential counselling for staff should surely be considered, in addition to a greater attention to dealing with stress at work such as is found in many areas of industry. It is ironic that an industry whose business is health pays so little attention to the well-being of its employees. Counselling for injured patients will also often be necessary, with the counsellor requiring a knowledge of medicine and medical practice as well as the knowledge and skills required for the treatment of trauma and depression.

The true costs of accidents and substandard care are unknown, but probably vastly greater than the obvious costs of litigation and compensation. The question is not can we afford safety programmes, but how can we not afford them?

Index

Note: Figures and Tables are indicated by *italic page numbers*, footnotes by suffix 'n'